QUASILINEAR CONTROL

Performance Analysis and Design of Feedback Systems with Nonlinear Sensors and Actuators

This is a textbook on quasilinear control (QLC). QLC is a set of methods for performance analysis and design of linear plant/nonlinear instrumentation (LPNI) systems. The approach of QLC is based on the method of stochastic linearization, which reduces the nonlinearities of actuators and sensors to quasilinear gains. Unlike the usual – Jacobian linearization – stochastic linearization is global. Using this approximation, QLC extends most of the linear control theory techniques to LPNI systems. In addition, QLC includes new problems, specific for the LPNI scenario. Examples include instrumented LQR/LQG, in which the controller is designed simultaneously with the actuator and sensor, and partial and complete performance recovery, in which the degradation of linear performance is either contained by selecting the right instrumentation or completely eliminated by the controller boosting.

ShiNung Ching is a Postdoctoral Fellow at the Neurosciences Statistics Research Laboratory at MIT, since completing his Ph.D. in electrical engineering at the University of Michigan. His research involves a systems theoretic approach to anesthesia and neuroscience, looking to use mathematical techniques and engineering approaches – such as dynamical systems, modeling, signal processing, and control theory – to offer new insights into the mechanisms of the brain.

Yongsoon Eun is a Senior Research Scientist at Xerox Innovation Group in Webster, New York. Since 2003, he has worked on a number of subsystem technologies in the xerographic marking process and image registration technology for the inkjet marking process. His interests are control systems with nonlinear sensors and actuators, cyclic systems, and the impact of multitasking individuals on organizational productivity.

Cevat Gokcek was an Assistant Professor of Mechanical Engineering at Michigan State University. His research in the Controls and Mechatronics Laboratory focused on automotive, aerospace, and wireless applications, with current projects in plasma ignition systems and resonance-seeking control systems to improve combustion and fuel efficiency.

Pierre T. Kabamba is a Professor of Aerospace Engineering at the University of Michigan. His research interests are in the area of linear and nonlinear dynamic systems, robust control, guidance and navigation, and intelligent control. His recent research activities are aimed at the development of a quasilinear control theory that is applicable to linear plants with nonlinear sensors or actuators. He has also done work in the design, scheduling, and operation of multi-spacecraft interferometric imaging systems, in analysis and optimization of random search algorithms, and in simultaneous path planning and communication scheduling for UAVs under the constraint of radar avoidance. He has more than 170 publications in refereed journals and conferences and numerous book chapters.

Semyon M. Meerkov is a Professor of Electrical Engineering at the University of Michigan. He received his Ph.D. from the Institute of Control Sciences in Moscow, where he remained until 1977. He then moved to the Department of Electrical and Computer Engineering at the Illinois Institute of Technology and to Michigan in 1984. He has held visiting positions at UCLA (1978–1979); Stanford University (1991); Technion, Israel (1997–1998 and 2008); and Tsinghua, China (2008). He was the editor-in-chief of *Mathematical Problems in Engineering*, department editor for *Manufacturing Systems of IIE Transactions*, and associate editor of several other journals. His research interests are in systems and control with applications to production systems, communication networks, and the theory of rational behavior. He is a Life Fellow of IEEE. He is the author of numerous research publications and books, including *Production Systems Engineering* (with Jingshang Li, 2009).

Quasilinear Control

Performance Analysis and Design of Feedback
Systems with Nonlinear Sensors and Actuators

ShiNung Ching
Massachusetts Institute of Technology

Yongsoon Eun
Xerox Research Center Webster

Cevat Gokcek
Michigan State University

Pierre T. Kabamba
University of Michigan

Semyon M. Meerkov
University of Michigan

CAMBRIDGE
UNIVERSITY PRESS

CAMBRIDGE
UNIVERSITY PRESS

32 Avenue of the Americas, New York NY 10013-2473, USA

Cambridge University Press is part of the University of Cambridge.

It furthers the University's mission by disseminating knowledge in the pursuit of education, learning and research at the highest international levels of excellence.

www.cambridge.org
Information on this title: www.cambridge.org/9781107429383

First published 2011
First paperback edition 2014

A catalogue record for this publication is available from the British Library

Library of Congress Cataloguing in Publication data
Quasilinear control : performance analysis and design of feedback systems
  with nonlinear sensors and actuators / ShiNung Ching ... [et al.].
     p. cm.
  Includes bibliographical references and index.
  ISBN 978-1-107-00056-8 (hardback)
  1. Stochastic control theory.   2. Quasilinearization.   I. Ching, ShiNung.
  QA402.37.Q37 2010
  629.8'312–dc22   2010039407

ISBN 978-1-107-00056-8 Hardback
ISBN 978-1-107-42938-3 Paperback

To my parents, with love,

SHINUNG CHING

To my wife Haengju, my son David, and my mother Ahn Young, with love and gratitude,

YONGSOON EUN

To my family, with love and gratitude,

PIERRE T. KABAMBA

To my dear wife Terry and to our children, Meera, Meir, Leah, and Rachel, with deepest love and admiration,

SEMYON M. MEERKOV

Brief Contents

Contents

Preface

Purpose: This volume is devoted to the study of feedback control of so-called *linear plant/nonlinear instrumentation* (LPNI) systems. Such systems appear naturally in situations where the plant can be viewed as linear but the instrumentation, that is, actuators and sensors, can not. For instance, when a feedback system operates effectively and maintains the plant close to a desired operating point, the plant may be linearized, but the instrumentation may not, because to counteract large perturbations or to track large reference signals, the actuator may saturate and the nonlinearities in sensors, for example, quantization and dead zones, may be activated.

The problems of stability and oscillations in LPNI systems have been studied for a long time. Indeed, the theory of absolute stability and the harmonic balance method are among the best known topics of control theory. More recent literature has also addressed LPNI scenarios, largely from the point of view of stability and anti-windup. However, the problems of performance analysis and design, for example, reference tracking and disturbance rejection, have not been investigated in sufficient detail. This volume is intended to contribute to this end by providing methods for designing *linear controllers* that ensure the desired *performance* of closed loop LPNI systems.

The methods developed in this work are similar to the usual linear system techniques, for example, root locus, LQR, and LQG, modified appropriately to account for instrumentation nonlinearities. Therefore, we refer to these methods as *quasilinear* and to the resulting area of control as *quasilinear control*.

Intent and prerequisites: This volume is intended as a textbook for a graduate course on quasilinear control or as a supplementary textbook for standard graduate courses on linear and nonlinear control. In addition, it can be used for self-study by practicing engineers involved in the analysis and design of control systems with nonlinear instrumentation.

The prerequisites include material on linear and nonlinear systems and control. Some familiarity with elementary probability theory and random processes may also be useful.

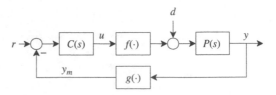

Figure 0.1. Linear plant/nonlinear instrumentation control system

Problems addressed: Consider the single-input single-output (SISO) system shown in Figure 0.1, where $P(s)$ and $C(s)$ are the transfer functions of the plant and the controller; $f(\cdot)$, $g(\cdot)$ are static odd nonlinearities characterizing the actuator and the sensor; and r, d, u, y, and y_m are the reference, disturbance, control, plant output, and sensor output, respectively. In the framework of this system and its multiple-input multiple-output (MIMO) generalizations, this volume considers the following problems:

P1. *Performance analysis:* Given $P(s)$, $C(s)$, $f(\cdot)$, and $g(\cdot)$, quantify the quality of reference tracking and disturbance rejection.

P2. *Narrow sense design:* Given $P(s)$, $f(\cdot)$, and $g(\cdot)$, design a controller $C(s)$ so that the quality of reference tracking and disturbance rejection meets specifications.

P3. *Wide sense design:* Given $P(s)$, design a controller $C(s)$ and select instrumentation $f(\cdot)$ and $g(\cdot)$ so that the quality of reference tracking and disturbance rejection meets specifications.

P4. *Partial performance recovery:* Let $C_\ell(s)$ be a controller, which is designed under the assumption that the actuator and the sensor are linear and which meets reference tracking and disturbance rejection specifications. Given $C_\ell(s)$, select $f(\cdot)$ and $g(\cdot)$ so that the performance degradation is guaranteed to be less than a given bound.

P5. *Complete performance recovery:* Given $f(\cdot)$ and $g(\cdot)$, modify, if possible, $C_\ell(s)$ so that performance degradation does not take place.

This volume provides conditions under which solutions of these problems exist and derives equations and algorithms that can be used to calculate these solutions.

Nonlinearities considered: We consider actuators and sensors characterized by piecewise continuous odd scalar functions. For example, we address:

• saturating actuators,

$$f(u) = \mathrm{sat}_\alpha(u) := \begin{cases} \alpha, & u > +\alpha, \\ u, & -\alpha \le u \le \alpha, \\ -\alpha, & u < -\alpha, \end{cases} \tag{0.1}$$

where α is the actuator authority;

- quantized sensors,

$$g(y) = \mathrm{qn}_\Delta(y) := \begin{cases} +\Delta \lfloor +y/\Delta \rfloor, & y \geq 0, \\ -\Delta \lfloor -y/\Delta \rfloor, & y < 0, \end{cases} \qquad (0.2)$$

where Δ is the quantization interval and $\lfloor u \rfloor$ denotes the largest integer less than or equal to y;

- sensors with a deadzone,

$$g(y) = \mathrm{dz}_\Delta(y) := \begin{cases} y - \Delta, & y > +\Delta, \\ 0, & -\Delta \leq u \leq +\Delta, \\ y + \Delta, & y < -\Delta, \end{cases} \qquad (0.3)$$

where 2Δ is the deadzone width.

The methods developed here are *modular* in the sense that they can be modified to account for any odd instrumentation nonlinearity just by replacing the general function representing the nonlinearity by a specific one corresponding to the actuator or sensor in question.

Main difficulty: LPNI systems are described by relatively complex nonlinear differential equations. Unfortunately, these equations cannot be treated by the methods of modern nonlinear control theory since the latter assumes that the control signal enters the state space equations in a linear manner and, thus, saturation and other nonlinearities are excluded. Therefore, a different approach to treat LPNI control systems is necessary.

Approach: The approach of this volume is based on the method of *stochastic linearization*, which is applicable to dynamical systems with random exogenous signals. Thus, we assume throughout this volume that both references and disturbances are random. However, several results on tracking deterministic references (e.g., step, ramp) are also included.

According to stochastic linearization, the static nonlinearities are replaced by *equivalent* or *quasilinear* gains N_a and N_s (see Figure 0.2, where $\hat{u}, \hat{y}$, and $\hat{y}_m$ replace u, y, and y_m). Unlike the usual Jacobian linearization, the resulting approximation is global, that is, it approximates the original system not only for small but for large signals as well. The price to pay is that the gains N_a and N_s depend not only on the nonlinearities $f(\cdot)$ and $g(\cdot)$, but also on all other elements of Figure 0.1, including the transfer functions and the exogenous signals, since, as it turns out, N_a and N_s are functions of the standard deviations, $\sigma_{\hat{u}}$ and $\sigma_{\hat{y}}$, of $\hat{u}$ and $\hat{y}$, respectively, that is, $N_a = N_a(\sigma_{\hat{u}})$ and $N_s = N_s(\sigma_{\hat{y}})$. Therefore, we refer to the system of Figure 0.2 as a *quasilinear control system*. Systems of this type are the main topic of study in this volume.

Thus, instead of assuming that a linear system represents the reality, as in linear control, we assume that a quasilinear system represents the reality and carry out

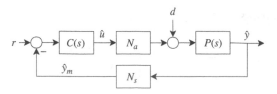

Figure 0.2. Quasilinear control system

control-theoretic developments of problems P1–P5, which parallel those of linear control theory, leading to what we call *quasilinear control* (QLC) *theory*.

The question of accuracy of stochastic linearization, that is, the precision with which the system of Figure 0.2 approximates that of Figure 0.1, is clearly of importance. Unfortunately, no general results in this area are available. However, various numerical and analytical studies indicate that if the plant, $P(s)$, is low-pass filtering, the approximation is well within 10% in terms of the variances of y and $\hat{y}$ and u and $\hat{u}$. More details on stochastic linearization and its accuracy are included in Chapter 2. It should be noted that stochastic linearization is somewhat similar to the method of harmonic balance, with $N_a(\sigma_{\hat{u}})$ and $N_s(\sigma_{\hat{y}})$ playing the roles of describing functions.

Book organization: The book consists of eight chapters. Chapter 1 places LPNI systems and quasilinear control in the general field of control theory. Chapter 2 describes the method of stochastic linearization as it applies to LPNI systems and derives equations for quasilinear gains in the problems of reference tracking and disturbance rejection. Chapters 3 and 4 are devoted to analysis of quasilinear control systems from the point of view of reference tracking and disturbance rejection, respectively (problem P1). Chapters 5 and 6 also address tracking and disturbance rejection problems, but from the point of view of design; both wide and narrow sense design problems are considered (problems P2 and P3). Chapter 7 addresses the issues of performance recovery (problems P4 and P5). Finally, Chapter 8 includes the proofs of all formal statements included in the book.

Each chapter begins with a short motivation and overview and concludes with a summary and annotated bibliography. Chapters 2–7 also include homework problems.

Acknowledgments: The authors thankfully acknowledge the stimulating environment at the University of Michigan, which was conducive to the research that led to this book. Financial support was provided for more than fifteen years by the National Science Foundation; gratitude to the Division of Civil, Mechanical and Manufacturing Innovations is in order.

Thanks are due to the University of Michigan graduate students who took a course based on this book and provided valuable comments: these include M.S. Holzel, C.T. Orlowski, H.-R. Ossareh, H.W. Park, H.A. Poonawala, and E.D. Summer. Special thanks are due to Hamid-Reza Ossareh, who carefully read every chapter of the manuscript and made numerous valuable suggestions. Also, the authors are grateful to University of Michigan graduate student Chris Takahashi, who participated in developing the LMI approach to LPNI systems.

The authors are also grateful to Peter Gordon, Senior Editor at Cambridge University Press, for his support during the last year of this project.

Needless to say, however, all errors, which are undoubtedly present in the book, are due to the authors alone. The list of corrections is maintained at http://www.eecs.umich.edu/~smm/monographs/QLC/.

Last, but not least, we are indebted to our families for their love and support, which made this book a reality.

1 Introduction

Motivation: This chapter is intended to introduce the class of systems addressed in this volume – the so-called Linear Plant/Nonlinear Instrumentation (LPNI) systems – and to characterize the control methodology developed in this book – Quasilinear Control (QLC).

Overview: After introducing the notions of LPNI systems and QLC and listing the problems addressed, the main technique of this book – the method of stochastic linearization – is briefly described and compared with the usual, Jacobian, linearization. In the framework of this comparison, it is shown that the former provides a more accurate description of LPNI systems than the latter, and the controllers designed using the QLC result, generically, yield better performance than those designed using linear control (LC). Finally, the content of the book is outlined.

1.1 Linear Plant/Nonlinear Instrumentation Systems and Quasilinear Control

Every control system contains nonlinear instrumentation – actuators and sensors. Indeed, the actuators are ubiquitously saturating; the sensors are often quantized; deadzone, friction, hysteresis, and so on are also encountered in actuator and sensor behavior.

Typically, the plants in control systems are nonlinear as well. However, if a control system operates effectively, that is, maintains its operation in a desired regime, the plant may be linearized and viewed as locally linear. The instrumentation, however, can not: to reject large disturbances, to respond to initial conditions sufficiently far away from the operating point, or to track large changes in reference signals – all may activate essential nonlinearities in actuators and sensors, resulting in fundamentally nonlinear behavior. These arguments lead to a class of systems that we refer to as *Linear Plant/Nonlinear Instrumentation* (LPNI).

The controllers in feedback systems are often designed to be linear. The main design techniques are based on root locus, sensitivity functions, LQR/LQG, H_∞, and so on, all leading to linear feedback. Although for LPNI systems both linear and

1

nonlinear controllers may be considered, to transfer the above-mentioned techniques to the LPNI case, we are interested in designing *linear* controllers. This leads to *closed loop LPNI systems*.

This volume is devoted to methods for analysis and design of closed loop LPNI systems. As it turns out, these methods are quite similar to those in the linear case. For example, root locus can be extended to LPNI systems, and so can LQR/LQG, H_∞, and so on. In each of them, the analysis and synthesis equations remain practically the same as in the linear case but coupled with additional transcendental equations, which account for the nonlinearities. That is why we refer to the resulting methods as *Quasilinear Control* (QLC) *Theory*. Since the main analysis and design techniques of QLC are not too different from the well-known linear control theoretic methods, QLC can be viewed as a simple addition to the standard toolbox of control engineering practitioners and students alike.

Although the term "LPNI systems" may be new, such systems have been considered in the literature for more than 50 years. Indeed, the theory of absolute stability was developed precisely to address the issue of global asymptotic stability of linear plants with linear controllers and sector-bounded actuators. For the same class of systems, the method of harmonic balance/describing functions was developed to provide a tool for limit cycle analysis. In addition, the problem of stability of systems with saturating actuators has been addressed in numerous publications. However, the issues of performance, that is, disturbance rejection and reference tracking, have been addressed to a much lesser extent. These are precisely the issues considered in this volume and, therefore, we use the subtitle *Performance Analysis and Design of Feedback Systems with Nonlinear Actuators and Sensors*.

In view of the above, one may ask a question: If all feedback systems include nonlinear instrumentation, how have controllers been designed in the past, leading to a plethora of successful applications in every branch of modern technology? The answer can be given as follows: In practice, most control systems are, indeed, designed ignoring the actuator and sensor nonlinearities. Then, the resulting closed loop systems are evaluated by computer simulations, which include nonlinear instrumentation, and the controller gains are readjusted so that the nonlinearities are not activated. Typically, this leads to performance degradation. If the performance degradation is not acceptable, sensors and actuators with larger linear domains are employed, and the process is repeated anew. This approach works well in most cases, but not in all: the Chernobyl nuclear accident and the crash of a YF-22 airplane are examples of its failures. Even when this approach does work, it requires a lengthy and expensive process of simulation and design/redesign. In addition, designing controllers so that the nonlinearities are not activated (e.g., actuator saturation is avoided) leads, as is shown in this book, to performance losses. Thus, developing methods in which the instrumentation nonlinearities are taken into account from the very beginning of the design process, is of significant practical importance. The authors of this volume have been developing such methods for more than 15 years, and the results are summarized in this volume.

As a conclusion for this section, it should be pointed out that modern Nonlinear Control Theory is not applicable to LPNI systems because it assumes that the control signals enter the system equations in a linear manner, thereby excluding saturation and other nonlinearities in actuators. Model Predictive Control may also be undesirable, because it is computationally extensive and, therefore, complex in implementation.

1.2 QLC Problems

Consider the closed loop LPNI system shown in Figure 1.1. Here the transfer functions $P(s)$ and $C(s)$ represent the plant and controller, respectively, and the nonlinear functions $f(\cdot)$ and $g(\cdot)$ describe, respectively, the actuator and sensor. The signals r, d, e, u, v, y, and y_m are the reference, disturbance, error, controller output, actuator output, plant output, and measured output, respectively. These notations are used throughout this book. In the framework of the system of Figure 1.1, this volume considers the following problems (rigorous formulations are given in subsequent chapters):

P1. **Performance analysis:** Given $P(s)$, $C(s)$, $f(\cdot)$, and $g(\cdot)$, quantify the performance of the closed loop LPNI system from the point of view of reference tracking and disturbance rejection.

P2. **Narrow sense design:** Given $P(s)$, $f(\cdot)$, and $g(\cdot)$, design, if possible, a controller so that the closed loop LPNI system satisfies the required performance specifications.

P3. **Wide sense design:** Given $P(s)$, design a controller $C(s)$ and select the instrumentation $f(\cdot)$ and $g(\cdot)$ so that the closed loop LPNI system satisfies the required performance specifications.

P4. **Partial performance recovery:** Assume that a controller, $C_l(s)$, is designed so that the closed loop system meets the performance specifications if the actuator and sensor were linear. Select $f(\cdot)$ and $g(\cdot)$ so that the performance degradation of the closed loop LPNI system with $C_l(s)$ does not exceed a given bound, as compared with the linear case.

P5. **Complete performance recovery:** As in the previous problem, let $C_l(s)$ be a controller that satisfies the performance specifications of the closed loop system with linear instrumentation. For given $f(\cdot)$ and $g(\cdot)$, redesign $C_l(s)$ so that the closed loop LPNI exhibits, if possible, no performance degradation.

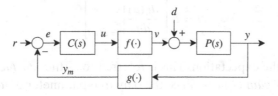

Figure 1.1. Closed loop LPNI system.

The first two of the above problems are standard in control theory, but are considered here for the LPNI case. The last three problems are specific to LPNI systems and have not been considered in linear control (LC). Note that the last problem is reminiscent of anti-windup control, whereby $C_l(s)$ is augmented by a mechanism that prevents the so-called windup of integral controllers in systems with saturating actuators.

1.3 QLC Approach: Stochastic Linearization

The approach of QLC is based on a quasilinearization technique referred to as stochastic linearization. This method was developed more than 50 years ago and since then has been applied in numerous engineering fields. Applications to feedback control have also been reported. However, comprehensive development of a control theory based on this approach has not previously been carried out. This is done in this volume.

Stochastic linearization requires exogenous signals (i.e., references and disturbances) to be random. While this is often the case for disturbances, the references are assumed in LC to be deterministic – steps, ramps, or parabolic signals. Are these the only references encountered in practice? The answer is definitely in the negative: in many applications, the reference signals can be more readily modeled as random than as steps, ramps, and so on. For example, in the hard disk drive control problem, the read/write head in both track-seeking and track-following operations is affected by reference signals that are well modeled by Gaussian colored processes. Similarly, the aircraft homing problem can be viewed as a problem with random references. Many other examples of this nature can be given. Thus, along with disturbances, QLC assumes that the reference signals are random processes and, using stochastic linearization, provides methods for designing controllers for both reference tracking and disturbance rejection problems. The standard, deterministic, reference signals are also used, for example, to develop the notion of LPNI system types and to define and analyze the notion of the so-called trackable domain.

The essence of stochastic linearization can be characterized as follows: Assume that the actuator is described by an odd piecewise differentiable function $f(u(t))$, where $u(t)$ is the output of the controller, which is assumed to be a zero-mean wide sense stationary (wss) Gaussian process. Consider the problem: approximate $f(u(t))$ by $Nu(t)$, where N is a constant, so that the mean-square error is minimized. It turns out (see Chapter 2) that such an N is given by

$$N = E\left[\left.\frac{df(u)}{du}\right|_{u=u(t)}\right],\qquad(1.1)$$

where E denotes the expectation. This is referred to as the *stochastically linearized gain* or *quasilinear gain* of $f(u)$. Since the only free parameter of $u(t)$ is its standard deviation, σ_u, it follows from (1.1) that the stochastically linearized gain depends on a single variable – the standard deviation of its argument; thus,

$$N = N(\sigma_u). \tag{1.2}$$

Note that stochastic linearization is indeed a quasilinear, rather than linear, operation: the quasilinear gains of $\alpha f(\cdot)$ and $f(\cdot)\alpha$, where α is a constant, are not the same, the former being $\alpha N(\sigma_u)$ the latter being $N(\alpha \sigma_u)$.

In the closed loop environment, σ_u depends not only on $f(u)$ but also on all other components of the system (i.e., the plant and the controller parameters) and on all exogenous signals (i.e., references and disturbances). This leads to transcendental equations that define the quasilinear gains. The study of these equations in the framework of various control-theoretic problems (e.g., root locus, sensitivity functions, LQR/LQG, H_∞) is the essence of the theory of QLC.

As in the open loop case, a stochastically linearized closed loop system is also not linear: its output to the sum of two exogenous signals is not equal to the sum of the outputs to each of these signals, that is, superposition does not hold. However, since, when all signals and functional blocks are given, the system has a constant gain N, we refer to a stochastically linearized closed loop system as *quasilinear*.

1.4 Quasilinear versus Linear Control

Consider the closed-loop LPNI system shown in Figure 1.2(a). If the usual Jacobian linearization is used, this system is reduced to that shown in Figure 1.2(b), where all signals are denoted by the same symbols as in Figure 1.2(a) but with a ~. In this

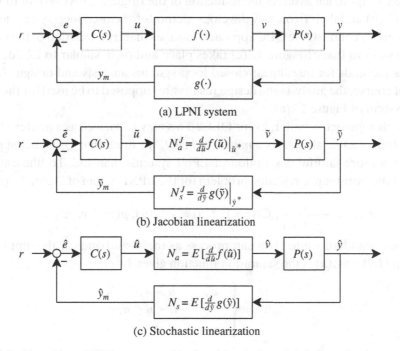

(a) LPNI system

(b) Jacobian linearization

(c) Stochastic linearization

Figure 1.2. Closed loop LPNI system and its Jacobian and stochastic linearizations.

system, the actuator and sensor are represented by constant gains evaluated as the derivatives of $f(\cdot)$ and $g(\cdot)$ at the operating point:

$$N_a^J = \left.\frac{df(\tilde{u})}{d\tilde{u}}\right|_{\tilde{u}=\tilde{u}*}, \tag{1.3}$$

$$N_s^J = \left.\frac{dg(\tilde{y})}{d\tilde{y}}\right|_{\tilde{y}=\tilde{y}*}. \tag{1.4}$$

Clearly, this system describes the original LPNI system of Figure 1.2(a) only locally, around the fixed operating point.

If stochastic linearization is used, the system of Figure 1.2(a) is reduced to the quasilinear one shown in Figure 1.2(c), where all signals are again denoted by the same symbols as in Figure 1.2(a) but with a ˆ; these notations are used throughout this book. As it is indicated above and discussed in detail in Chapter 2, here the actuator and sensor are represented by their quasilinear gains:

$$N_a(\sigma_{\hat{u}}) = E\left[\frac{df(\hat{u})}{d\hat{u}}|_{\hat{u}=\hat{u}(t)}\right], \tag{1.5}$$

$$N_s(\sigma_{\hat{y}}) = E\left[\frac{dg(\hat{y})}{d\hat{y}}|_{\hat{y}=\hat{y}(t)}\right]. \tag{1.6}$$

Since $N_a(\sigma_{\hat{u}})$ and $N_s(\sigma_{\hat{y}})$ depend not only on $f(\cdot)$ and $g(\cdot)$ but also on all elements of the system in Figure 1.2(c), the quasilinearization describes the closed loop LPNI system globally, with "weights" defined by the statistics of $\hat{u}(t)$ and $\hat{y}(t)$.

The LC approach assumes the reduction of the original LPNI system to that of Figure 1.2(b) and then rigorously develops methods for closed loop system analysis and design. In contrast, the QLC approach assumes that the reduction of the original LPNI system to that of Figure 1.2(c) takes place and then, similar to LC, develops rigorous methods for quasilinear closed loop systems analysis and design. In both cases, of course, the analysis and design results are supposed to be used for the actual LPNI system of Figure 1.2(a).

Which approach is better, LC or QLC? This may be viewed as a matter of belief or a matter of calculations. *As a matter of belief*, we think that QLC, being global, provides a more faithful description of LPNI systems than LC. To illustrate this, consider the disturbance rejection problem for the LPNI system of Figure 1.2(a) with

$$P(s) = \frac{1}{s^2+s+1}, \ C(s) = 1, \ f(u) = \text{sat}_\alpha(u), \ g(y) = y, \ r(t) = 0 \tag{1.7}$$

and with a standard white Gaussian process as the disturbance at the input of the plant. In (1.7), $\text{sat}_\alpha(u)$ is the saturation function given by

$$\text{sat}_\alpha(u) = \begin{cases} \alpha, & u > +\alpha, \\ u, & -\alpha \leq u \leq \alpha, \\ -\alpha, & u < -\alpha. \end{cases} \tag{1.8}$$

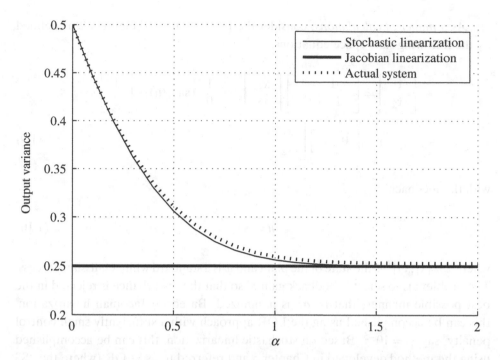

Figure 1.3. Comparison of stochastic linearization, Jacobian linearization, and actual system performance.

For this LPNI system, we construct its Jacobian and stochastic linearizations and calculate the variances, $\sigma_{\tilde{y}}^2$ and $\sigma_{\hat{y}}^2$, of the outputs $\tilde{y}(t)$ and $\hat{y}(t)$ as functions of α. (Note that $\sigma_{\tilde{y}}^2$ is calculated using the usual Lyapunov equation approach and $\sigma_{\hat{y}}^2$ is calculated using the stochastic linearization approach developed in Chapter 2.) In addition, we simulate the actual LPNI system of Figure 1.2(a) and numerically evaluate σ_y^2. All three curves are shown in Figure 1.3. From this figure, we observe the following:

- The Jacobian linearization of $\mathrm{sat}_\alpha(u)$ is independent of α, thus, the predicted variance is constant.
- When α is large (i.e., the input is not saturated), Jacobian linearization is accurate. However, it is highly inaccurate for small values of α.
- Stochastic linearization accounts for the nonlinearity and, thus, predicts an output variance that depends on α.
- Stochastic linearization accurately matches the actual performance for *all* values of α.

We believe that a similar situation takes place for any closed loop LPNI system: Stochastic linearization, when applicable, describes the actual LPNI system more faithfully than Jacobian linearization. (As shown in Chapter 2, stochastic linearization is applicable when the plant is low-pass filtering.)

As a matter of calculations, consider the LPNI system of Figure 1.2(a) defined by the following state space equations:

$$\begin{bmatrix} \dot{x}_1 \\ \dot{x}_2 \end{bmatrix} = \begin{bmatrix} -1 & -1 \\ 1 & 0 \end{bmatrix} \begin{bmatrix} x_1 \\ x_2 \end{bmatrix} + \begin{bmatrix} 1 \\ 0 \end{bmatrix} \mathrm{sat}_\alpha(u) + \begin{bmatrix} 1 \\ 0 \end{bmatrix} w$$

$$y = \begin{bmatrix} 0 & 1 \end{bmatrix} \begin{bmatrix} x_1 \\ x_2 \end{bmatrix}, \tag{1.9}$$

with the feedback

$$u = Kx, \tag{1.10}$$

where $x = [x_1, x_2]^T$ is the state of the plant and w is a standard white Gaussian process. The problem is to select a feedback gain K so that the disturbance is rejected in the best possible manner, that is, σ_y^2 is minimized. Based on Jacobian linearization, this can be accomplished using the LQR approach with a sufficiently small control penalty, say, $\rho = 10^{-5}$. Based on stochastic linearization, this can be accomplished using the method developed in Chapter 5 and referred to as SLQR (where the "S" stands for "saturating") with the same ρ. The resulting controllers, of course, are used in the LPNI system. Simulating this system with the LQR controller and with the SLQR controller, we evaluated numerically σ_y^2 for both cases. The results are shown in Figure 1.4 as a function of the saturation level. From this figure, we conclude the following:

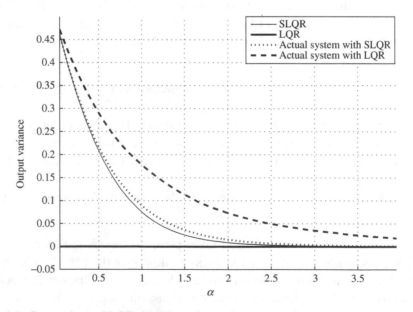

Figure 1.4. Comparison of LQR, SLQR, and actual system performance.

- Since ρ is small and the plant is minimum phase, LQR provides a high gain solution that renders the output variance close to zero. Due to the underlying Jacobian linearization, this solution is constant for all α.
- Due to the input saturation, the performance of the actual system with the LQR controller is significantly worse than the LQR design, even for relatively large values of α.
- The SLQR solution explicitly accounts for α and, thus, yields a nonzero output variance.
- The performance of the actual system with an SLQR controller closely matches the intended design.
- The actual SLQR performance exceeds the actual LQR performance for all values of α.

As shown, using LQR in this situation is deceiving since the actual system can never approach the intended performance. In contrast, the SLQR solution is highly representative of the actual system behavior (and, indeed, exceeds the actual LQR performance). In fact, it is possible to prove that QLC-based controllers (e.g., controllers designed using SLQR) generically ensure better performance of LPNI systems than LC-based controllers (e.g., based on LQR).

These comparisons, we believe, justify the development and utilization of QLC.

1.5 Overview of Main QLC Results

This section outlines the main QLC results included in this volume.

Chapter 2 describes the method of stochastic linearization in the framework of LPNI systems. After deriving the expression for quasilinear gain (1.1) and illustrating it for typical nonlinearities of actuators and sensors, it concentrates on closed loop LPNI systems (Figure 1.2(a)) and their stochastic linearizations (Figure 1.2(c)). Since the quasilinear gain of an actuator, N_a, depends on the standard deviation of the signal at its input, $\sigma_{\hat{u}}$ and, in turn, $\sigma_{\hat{u}}$ depends on N_a, the quasilinear gain of the actuator is defined by a *transcendental equation*. The same holds for the quasilinear gain of the sensor. Chapter 2 derives these transcendental equations for various scenarios of reference tracking and disturbance rejection. For instance, in the problem of reference tracking with a nonlinear actuator and linear sensor, the quasilinear gain of the actuator is defined by the equation

$$N_a = \mathcal{F}\left(\left\| \frac{F_{\Omega_r}(s)C(s)}{1 + P(s)N_aC(s)} \right\|_2\right), \tag{1.11}$$

where

$$\mathcal{F}(\sigma) = \int\limits_{-\infty}^{\infty} \left[\frac{d}{dx}f(x)\right] \frac{1}{\sqrt{2\pi}\sigma} \exp\left(-\frac{x^2}{2\sigma^2}\right) dx. \tag{1.12}$$

Here, $F_{\Omega_r}(s)$ is the reference coloring filter, $f(x)$ is the nonlinear function that describes the actuator, and $\|\cdot\|_2$ is the 2-norm of a transfer function. Chapter 2 provides a sufficient condition under which this and similar equations for other performance problems have solutions and formulates a bisection algorithm to find them with any desired accuracy. Based on these solutions, the performance of closed loop LPNI systems in problems of reference tracking and disturbance rejection is investigated. Finally, Chapter 2 addresses the issue of accuracy of stochastic linearization and shows (using the *Fokker-Planck equation* and the *filter hypothesis*) that the error between the standard deviation of the plant output and its quasilinearization (i.e., σ_y and $\sigma_{\hat{y}}$) is well within 10%, if the plant is low-pass filtering. The equations derived in Chapter 2 are used throughout the book for various problems of performance analysis and design.

Chapter 3 is devoted to analysis of reference tracking in closed loop LPNI systems. Here, the notion of *system type* is extended to feedback control with saturating actuators, and it is shown that the type of the system is defined by the plant poles at the origin (rather than the loop transfer function poles at the origin, as it is in the linear case). The controller poles, however, also play a role, but a minor one compared with those of the plant. In addition, Chapter 3 introduces the notion of *trackable domains*, that is, the ranges of step, ramp, and parabolic signals that can be tracked by LPNI systems with saturating actuators. In particular, it shows that the trackable domain (TD) for step inputs, $r(t) = r_0\mathbf{1}(t)$, where r_0 is a constant and $\mathbf{1}(t)$ is the unit step function, is given by

$$TD = \{r_0 : |r_0| < \left|\frac{1}{C_0} + P_0\right|\alpha\}, \tag{1.13}$$

where C_0 and P_0 are d.c. gains of the controller and plant, respectively, and α is the level of actuator saturation. Thus, TD is finite, unless the plant has a pole at the origin.

While the above results address the issue of tracking deterministic signals, Chapter 3 investigates also the problem of random reference tracking. First, linear systems are addressed. As a motivation, it is shown that the standard deviation of the error signal, σ_e, is a poor predictor of tracking quality since for the same σ_e track loss can be qualitatively different. Based on this observation, the so-called *tracking quality indicators*, similar to gain and phase margins in linear systems, are introduced. The main instrument here is the so-called *random sensitivity function* (RS). In the case of linear systems, this function is defined by

$$RS(\Omega) = ||F_\Omega(s)S(s)||_2, \tag{1.14}$$

where, as before, $F_\Omega(s)$ is the reference signal coloring filter with 3dB bandwidth Ω and $S(s)$ is the usual sensitivity function. The bandwidth of $RS(\Omega)$, its d.c. gain, and the resonance peak define the tracking quality indicators, which are used as specifications for tracking controller design.

Finally, Chapter 3 transfers the above ideas to tracking random references in LPNI systems. This development is based on the so-called *saturating random sensitivity* function, $SRS(\Omega,\sigma_r)$, defined as

$$SRS(\Omega,\sigma_r) = \frac{RS(\Omega)}{\sigma_r}, \tag{1.15}$$

where $RS(\Omega)$ is the random sensitivity function of the stochastically linearized version of the LPNI system and σ_r is the standard deviation of the reference signal. Using $SRS(\Omega,\sigma_r)$, an additional tracking quality indicator is introduced, which accounts for the trackable domain and indicates when and to what extent amplitude truncation takes place. In conclusion, Chapter 3 presents a diagnostic flowchart that utilizes all tracking quality indicators to predict the tracking capabilities of LPNI systems with saturating actuators. These results, which transfer the frequency (ω) domain methods of LC to the frequency (Ω) domain methods of QLC, can be used for designing tracking controllers by shaping $SRS(\Omega,\sigma_r)$. The theoretical developments of Chapter 3 are illustrated using the problem of hard disk drive control.

Chapter 4 is devoted to analysis of the disturbance rejection problem in closed loop LPNI systems. Here, the results of Chapter 2 are extended to the multiple-input-multiple-output (MIMO) case. In addition, using an extension of the LMI approach, Chapter 4 investigates fundamental limitations on achievable disturbance rejection due to actuator saturation and shows that these limitations are similar to those imposed by non-minimum-phase zeros in linear systems. The final section of this chapter shows how the analysis of LPNI systems with rate saturation and with hysteresis can be reduced to the amplitude saturation case.

Chapter 5 addresses the issue of designing tracking controllers for LPNI systems in the time domain. The approach here is based on the so-called S-root locus, which is the extension of the classical root locus to systems with saturating actuators. This is carried out as follows: Consider the LPNI system of Figure 1.5(a) and its stochastic linearization of Figure 1.5(b). The saturated root locus of the system of Figure 1.5(a) is the path traced by the poles of the quasilinear system of Figure 1.5(b) when K changes from 0 to ∞. If N were independent of K, the S-root locus would coincide with the usual root locus. However, since $N(K)$ may tend to 0 as $K \to \infty$, the behavior of the S-root locus is defined by $\lim_{K\to\infty} KN(K)$. If this limit is infinite, the S-root locus coincides with the usual root locus. If this limit is finite, the S-root locus terminates prematurely, prior to reaching the open-loop zeros. These points

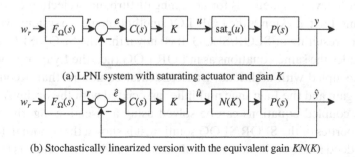

(a) LPNI system with saturating actuator and gain K

(b) Stochastically linearized version with the equivalent gain $KN(K)$

Figure 1.5. Systems for S-root locus design.

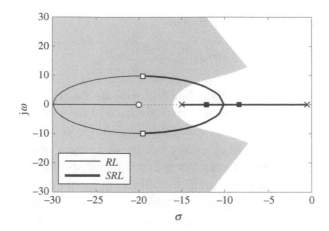

Figure 1.6. Saturated root locus.

are referred to as *termination points*, and Chapter 5 shows that they can be evaluated
using the positive solution, β^*, of the following equation:

$$\beta - \left\| \frac{F_\Omega(s)\,C(s)}{1 + \left(\frac{\alpha\sqrt{2/\pi}}{\beta}\right)P(s)\,C(s)} \right\|_2 = 0, \tag{1.16}$$

where, as before, $F_\Omega(s)$, $P(s)$, and $C(s)$ are the reference coloring filter, the plant and
the controller, respectively. An example of the S-root locus (SRL) and the classical
root locus (RL) is shown in Figure 1.6, where the termination points are indicated
by white squares, the shaded area is the admissible domain, defined by the tracking
quality indicators of Chapter 3, and the rest of the notations are the same as in the
classical root locus.

In addition, Chapter 5 introduces the notion of *truncation points*, which indicate the segments of the S-root locus corresponding to poles leading to amplitude
truncation. These points are shown in Figure 1.6 by black squares; all poles beyond
these locations result in loss of tracking due to truncations. To "push" the truncation
points in the admissible domain, the level of saturation must be necessarily increased.
These results provide an approach to tracking controller design for LPNI systems in
the time domain.

Chapter 6 develops methods for designing disturbance rejection controllers for
LPNI systems. First, it extends the LQR/LQG methodologies to systems with saturating actuators, resulting in SLQR/SLQG. It is shown that the SLQR/SLQG synthesis
engine includes the same equations as in LQR/LQG (i.e., the Lyapunov and Riccati
equations) coupled with additional transcendental equations that account for the
quasilinear gain and the Lagrange multiplier associated with the optimization problem. These coupled equations can be solved using a bisection algorithm. Among
various properties of the SLQR/SLQG solution, it is shown that optimal disturbance
rejection indeed requires the activation of saturation, which contradicts the intuitive
opinion that it should be avoided. So the question "to saturate or not to saturate" is

answered in the affirmative. Another technique developed in Chapter 6 is referred to as ILQR/ILQG, where the "I" stands for "Instrumented." The problem here is to design simultaneously the controller and the instrumentation (i.e., actuator and sensor) so that a performance index is optimized. The performance index is given by

$$J = \sigma_{\hat{y}}^2 + \rho \sigma_{\hat{u}}^2 + W(\alpha, \beta), \tag{1.17}$$

where $\rho > 0$ is the control penalty and W models the "cost" of the instrumentation as a function of the parameters α of the actuator and β of the sensor. Using the Lagrange multipliers approach, Chapter 6 provides a solution of this optimization problem, which again results in Lyapunov and Riccati equations coupled with transcendental relationships. The developments of Chapter 6 are illustrated by the problem of ship roll stabilization under sea wave disturbance modeled as a colored noise.

Chapter 7 is devoted to performance recovery in LPNI systems. The problems here are as follows: Let the controller, $C_l(s)$, be designed to satisfy performance specifications under the assumption that the actuator and sensor are linear. How should the parameters of the real, that is, nonlinear, actuator and sensor be selected so that the performance of the resulting LPNI system with the same $C_l(s)$ will not degrade below a given bound? This problem is referred to as *partial performance recovery*. The *complete performance recovery* problem is to redesign $C_l(s)$ so that the LPNI system exhibits the same performance as the linear one. The solution of the partial performance recovery problem is provided in terms of the Nyquist plot of the loop gain of the linear system. Based on this solution, the following rule of thumb is obtained: To ensure performance degradation of no more than 10%, the actuator saturation should be at least twice larger than the standard deviation of the controller output in the linear system, that is,

$$\alpha > 2\sigma_{u_l}. \tag{1.18}$$

The problem of complete performance recovery is addressed using the idea of boosting $C_l(s)$ gains to account for the drop in equivalent gains due to actuator and sensor nonlinearities. The so-called a- and s-boosting are considered, referring to boosting gains due to actuator and sensor nonlinearities, respectively. In particular, it is shown that a-boosting is possible if and only if the equation

$$x \mathcal{F}\left(x \left\| \frac{P(s)C(s)}{1 + P(s)C(s)} \right\|_2 \right) = 1 \tag{1.19}$$

with $\mathcal{F}$ defined in (1.12) has a positive solution. Based on this equation, the following rule of thumb is derived: Complete performance recovery in LPNI systems with saturating actuators is possible if

$$\alpha > 1.25\sigma_{u_l}, \tag{1.20}$$

where all notations are the same as in (1.18). Thus, if the level of actuator saturation satisfies (1.20), the linear controller can be boosted so that no performance degradation takes place. A method for finding the boosting gain is also provided. The validation of the boosting approach is illustrated using a magnetic levitation system.

The final chapter of the book, Chapter 8, provides the proofs of all formal statements included in the book.

As it follows from the above overview, this volume transfers most of LC to QLC. Specifically, the saturating random sensitivity function and the tracking quality indicators accomplish this for frequency domain techniques, the S-root locus for time domain techniques, and SLQR/SLQG for state space techniques. In addition, the LPNI-specific problems, for example, truncation points of the root locus, instrumentation selection, and the performance recovery, are also formulated and solved.

1.6 Summary

- The analysis and design of closed loop linear plant/nonlinear instrumentation (LPNI) systems is the main topic of this volume.
- The goal is to extend the main analysis and design techniques of linear control (LC) to the LPNI case. Therefore, the resulting methods are referred to as quasilinear control (QLC).
- The approach of QLC is based on the method of stochastic linearization. According to this method, an LPNI system is represented by a quasilinear one, where the static nonlinearities are replaced by the expected values of their gradients. As a result, stochastic linearization represents the LPNI system globally (rather than locally, as it is in the case of Jacobian linearization).
- Stochastic linearizations of LPNI systems represent the actual LPNI systems more faithfully than Jacobian linearization.
- Starting from stochastically linearized versions of LPNI systems, QLC develops methods for analysis and design that are as rigorous as those of LC (which starts from Jacobian linearization).
- This volume transfers most LC methods to QLC: The saturated random sensitivity function and the tracking quality indicators accomplish this for frequency domain techniques; S-root locus – for time domain techniques; and SLQR/SLQG – for state space techniques.
- In addition, several LPNI-specific problems, for example, truncation points of the root locus, instrumentation selection, and the performance recovery, are formulated and solved.

1.7 Annotated Bibliography

There is a plethora of monographs on design of linear feedback systems. Examples of undergraduate text are listed below:

[1.1] B.C. Kuo, *Automatic Control Systems*, Fifth Edition, Prentice Hall, Englewood Cliffs, NJ, 1987

[1.2] K. Ogata, *Modern Control Engineering*, Second Edition, Prentice Hall, Englewood Cliffs, NJ, 1990

[1.3] R.C. Dorf and R.H. Bishop, *Modern Control Systems*, Eighth Edition, Addison-Wesley, Menlo Park, CA, 1998

[1.4] G.C. Godwin, S.F. Graebe, and M.E. Salgado, *Control Systems Design*, Prentice Hall, Upper Shaddle River, NJ, 2001

[1.5] G.F. Franklin, J.D. Powel, and A. Emami-Naeini, *Feedback Control of Dynamic Systems*, Fourth Edition, Prentice Hall, Englewood Cliffs, NJ, 2002

At the graduate level, the following can be mentioned:

[1.6] I.M. Horowitz, *Synthesis of Feedback Systems*, Academic Press, London, 1963

[1.7] H. Kwakernaak and R. Sivan, *Linear Optimal Control Systems*, Wiley-Interscience, New York, 1972

[1.8] W.M. Wonham, *Linear Multivariable Control: A Geometric Approach*, Third Edition, Springer-Verlag, New York, 1985

[1.9] B.D.O. Anderson and J.B. Moore, *Optimal Control: Linear Quadratic Methods*, Prentice Hall, Englewood Cliffs, NJ, 1989

[1.10] J.M. Maciejowski, *Multivariable Feedback Design*, Addison-Wesley, Reading, MA, 1989

[1.11] K. Zhou, J.C. Doyle, and K. Glover, *Robust and Optimal Control*, Prentice Hall, Upper Saddle River, NJ, 1996

The theory of absolute stability has its origins in the following:

[1.12] A.I. Lurie and V.N. Postnikov, "On the theory of stability of control systems," *Applied Mathematics and Mechanics*, Vol. 8, No. 3, pp. 246–248, 1944 (in Russian)

[1.13] M.A. Aizerman, "On one problem related to 'stability-in-the-large' of dynamical systems," *Russian Mathematics Uspekhi*, Vol. 4, No. 4, pp. 187–188, 1949 (in Russian)

Subsequent developments are reported in the following:

[1.14] V.M. Popov, "On absolute stability of nonlinear automatic control systems," *Avtomatika i Telemekhanika*, No. 8, 1961 (in Russian). English translation: *Automation and Remote Control*, Vol. 22, No. 8, pp. 961–979, 1961

[1.15] V.A. Yakubovich, "The solution of certain matrix inequalities in automatic control theory," *Doklady Akademii Nauk*, Vol. 143, pp. 1304–1307, 1962 (in Russian)

[1.16] M.A. Aizerman and F.R. Gantmacher, *Absolute Stability of Regulator Systems*. Holden-Day, San Francisco, 1964 (Translated from the Russian original, *Akad. Nauk SSSR*, Moscow, 1963)

[1.17] R. Kalman, "Lyapunov functions for the problem of Lurie in automatic control, *Proc. of the National Academy of Sciences of the United States of America*, Vol. 49, pp. 201–205, 1963

[1.18] K.S. Narendra and J. Taylor, *Frequency Domain Methods for Absolute Stability*, Academic Press, New York, 1973

The method of harmonic balance has originated in the following:

[1.19] L.S. Goldfarb, "On some nonlinearities in regulator systems," *Avtomatika i Telemekhanika*, No. 5, pp. 149–183, 1947 (in Russian).

[1.20] R. Kochenburger, "A frequency response method for analyzing and synthesizing contactor servomechanisms," *Trans. AIEE*, Vol. 69, pp. 270–283, 1950

This was followed by several decades of further development and applications. A summary of early results can be found in the following:

[1.21] A. Gelb and W.E. Van der Velde, *Multiple-Input Describing Function and Nonlinear System Design*, McGraw-Hill, New York, 1968,

while later ones in

[1.22] A.I. Mees, "Describing functions – 10 years later," *IMA Journal of Applied Mathematics*, Vol. 32, No. 1–3, pp. 221–233, 1984

For the justification of this method (based on the idea of "filter hypothesis") and evaluation of its accuracy, see the following:

[1.23] M.A. Aizerman, "Physical foundations for application small parameter methods to problems of automatic control," *Avtomatika i Telemekhanika*, No. 5, pp. 597–603, 1953 (in Russian)

[1.24] E.M. Braverman, S.M. Meerkov, and E.S. Piatnitsky, "A small parameter in the problem of justifying the harmonic balance method (in the case of the filter hypothesis)," *Avtomatika i Telemekhanika*, No. 1, pp. 5–21, 1975 (in Russian). English translation: *Automation and Remote Control*, Vol. 36, No. 1, pp. 1–16, 1975

Using the notion of the mapping degree, this method has been justified in the following:

[1.25] A.R. Bergen and R.L. Frank, "Justification of the describing function method," *SIAM Journal of Control*, Vol. 9, No. 4, pp. 568–589, 1971

[1.26] A.I. Mees and A.R. Bergen, "Describing functions revisited," *IEEE Transactions on Automatic Control*, Vol. AC-20, No. 4, pp. 473–478, 1975

Several monographs that address the issue of stability of LPNI systems with saturating actuators can be found in the following:

[1.27] T. Hu and Z. Lin, *Control Systems with Actuator Saturation*, Birkauser, Boston, MA, 2001

[1.28] A. Saberi, A.A. Stoorvogel, and P. Sannuti, *Control of Linear Systems with Regulation and Input Constraints*, Springer-Verlag, New York, 2001

[1.29] V. Kapila and K.M. Grigoriadis, Ed., *Actuator Saturation Control*, Marcel Dekker, Inc., New York, 2002

Remarks on the saturating nature of the Chernobyl nuclear accident can be found in the following:

[1.30] G. Stein, "Respect for unstable," Hendrik W. Bode Lecture, *Proc. ACC*, Tampa, FL, 1989

Reasons for the crash of the YF-22 aircraft are reported in the following:

[1.31] M.A. Dornheim, "Report pinpoints factors leading to YF-22 crash", *Aviation Week & Space Technology.*, Vol. 137, No. 19, pp. 53–54, 1992

Modern theory of nonlinear control based on the geometric approach has its origin in the following:

[1.32] R.W. Brockett, "Asymptotic stability and feedback stabilization," in *Differential Geometric Control Theory*, R.W. Brockett, R.S. Millman, and H.J. Sussmann, Eds., pp. 181–191, 1983

Further developments are reported in the following:

[1.33] A. Isidori, *Nonlinear Control Systems*, Third Edition, Springer-Verlag, New York, 1995

Model predictive control was advanced in the following:

[1.34] J. Richalet, A. Rault, J.L. Testud, and J. Papon, "Model predictive heuristic control: Applications to industrial processes," *Automatica*, Vol. 14, No. 5, pp. 413–428, 1978

[1.35] C.R. Cutler and B.L. Ramaker, "Dynamic matrix control – A computer control algorithm," in *AIChE 86th National Meeting*, Houston, TX, 1979

Additional results can be found in

[1.36] C.E. Garcia, D.M. Prett, and M. Morari, "Model predictive control: Theory and practice – a survey," *Automatica*, Vol. 25, No. 3, pp. 338–349, 1989

[1.37] E.G. Gilbert and K. Tin Tan, "Linear systems with state and control constraints: the theory and applications of maximal output admissible sets," *IEEE Transactions Automatic Control*, Vol. 36, pp. 1008–1020, 1995

[1.38] D.Q. Mayne, J.B. Rawlings, C.V. Rao, and P.O.M. Scokaert, "Constrained model predictive control: Stability and optimality," *Automatica*, Vol. 36, pp. 789–814, 2000

[1.39] E.F. Camacho and C. Bordons, *Model Predictive Control*, Springer-Verlag, London, 2004

The term integrator "windup" seems to have appeared in

[1.40] J.C. Lozier, "A steady-state approach to the theory of saturable servo systems," *IRE Transactions on Automatic Control*, pp. 19–39, May 1956

Early work on antiwindup can be found in the following:

[1.41] H.A. Fertic and C.W. Ross, "Direct digital control algorithm with anti-windup feature," *ISA Transactions*, Vol. 6, No. 4, pp. 317–328, 1967

More recent results can be found in the following:

[1.42] M.V. Kothare, P.J. Campo, M. Morari, and C.N. Nett, "A unified framework for the study of anti-windup designs," *Automatica*, Vo. 30, No. 12, pp. 1869–1883, 1994

[1.43] N. Kapoor, A.R. Teel, and P. Daoutidis, "An anti-windup design for linear systems with input saturation," *Automatica*, Vol. 34, No. 5, pp. 559–574, 1998

[1.44] P. Hippe, *Windup in Control: Its Effects and Their Prevention*, Springer, London, 2006

The method of stochastic linearization originated in the following:

[1.45] R.C. Booton, M.V. Mathews, and W.W. Seifert, "Nonlinear servomechanisms with random inputs," *Dyn. Ana. Control Lab*, MIT, Cambridge, MA, 1953

[1.46] R.C. Booton, "The analysis of nonlinear systems with random inputs," *IRE Transactions on Circuit Theory*, Vol. 1, pp. 32–34, 1954

[1.47] I.E. Kazakov, "Approximate method for the statistical analysis of nonlinear systems," *Trudy VVIA* 394, 1954 (in Russian)

[1.48] I.E. Kazakov, "Approximate probability analysis of operational position of essentially nonlinear feedback control systems," *Automation and Remote Control*, Vol. 17, pp. 423–450, 1955

Various extensions can be found in the following:

[1.49] V.S. Pugachev, *Theory of Random Functions*, Pergamon Press, Elmsford, NY, 1965 (translation from Russian)

[1.50] I. Elishakoff, "Stoshastic linearization technique: A new interpretation and a selective review," *The Shock and Vibration Digest*, Vol. 32, pp. 179–188, 2000

[1.51] J.B. Roberts and P.D. Spanos, *Random Vibrations and Statistical Linearization*, Dover Publications, Inc., Mineola, NY, 2003

[1.52] L. Socha, *Linearization Methods for Stochastic Systems*, Springer, Berlin Heidelberg, 2008

Applications to control problems have been described in the following:

[1.53] I.E. Kazakov and B.G. Dostupov, *Statistical Dynamics of Nonlinear Control Systems*, Fizmatgiz, Moscow 1962 (In Russian)

[1.54] A.A. Pervozvansky, *Stochastic Processes in Nonlinear Control Systems*, Fizmatgiz, Moscow 1962 (in Russian)

[1.55] I.E. Kazakov, "Statistical analysis of systems with multi-dimensional nonlinearities," *Automation and Remote Control*, Vol. 26, pp. 458–464, 1965

and also in reference [1.21]
The stochastic modeling of reference signals in the problem of hard drive control can be found in the following:

[1.56] A. Silberschatz and P.B. Galvin, *Operating Systems Concepts*, Addison-Wesley, 1994

[1.57] T.B. Goh, Z. Li and B.M. Chen, "Design and implementation of a hard disk servo system using robust abd perfect tracking approach," *IEEE Transactions on Control Systems Technology*, Vol. 9, pp. 221–233, 2001

For the aircraft homing problem, similar conclusions can be deduced from the following:
[1.58] C.-F. Lin, *Modern Navigation, Guidance, and Control Processing*, Prentice Hall, Englewood Cliffs, NJ, 1991

[1.59] E.J. Ohlmeyer, "Root-mean-square miss distance of proportional navigation missile against sinusoidal target," *Journal of Guidance, Control and Dynamics*, Vol. 19, No. 3, pp. 563–568, 1996

In automotive problems, stochastic reference signals appear in the following:
[1.60] H.S. Bae and J.C. Gerdes, "Command modification using input shaping for automated highway systems with heavy trucks," *California PATH Research Report*, 1(UCB-ITS-PRR-2004-48), Berkeley, CA, 2004

The usual, Jacobian, linearization is the foundation of all methods for analysis and design on linear systems, including the indirect Lyapunov method. For more information see the following:
[1.61] M. Vidyasagar, *Nonlinear Systems Analysis*, Second Edition, Prentice Hall, Englewood Cliffs, NJ, 1993

[1.62] H.K. Khalil, *Nonlinear Systems*, Third Edition, Prentice Hall, Upper Saddle River, NJ, 2002

A discussion on calculating the 2-norm of a transfer function can be found in
[1.63] K. Zhou and J.C. Doyle, *Essentials of Robust Control*, Prentice Hall, Upper Saddle River, NJ, 1999

For the theory of Fokker-Planck equation turn to

[1.64] L. Arnold, *Stochastic Differential Equations*, Wiley Interscience, New York, 1973

[1.65] H. Risken, *The Fokker-Planck Equation: Theory and Applications*, Springer-Verlag, Berlin, 1989

[1.66] Z. Schuss, *Theory and Applications of Stochastic Processes*, Springer, New York, 2009

2 Stochastic Linearization of LPNI Systems

Motivation: This chapter is intended to present the main mathematical tool of this book – the method of stochastic linearization – in terms appropriate for the subsequent analysis and design of closed loop LPNI systems. Those familiar with this method are still advised to read this chapter since it derives equations used throughout this volume.

Overview: First, we present analytical expressions for the stochastically linearized (or quasilinear) gains of open loop systems. Then we derive transcendental equations that define the quasilinear gains of various types of closed loop LPNI systems. Finally, we discuss the accuracy of stochastic linearization in predicting the standard deviations of various signals in closed loop LPNI systems.

2.1 Stochastic Linearization of Open Loop Systems

2.1.1 Stochastic Linearization of Isolated Nonlinearities

Quasilinear gain: Consider Figure 2.1, where $f(u)$ is an odd piece wise differentiable function, $u(t)$ is a zero-mean wide sense stationary (wss) Gaussian process,

$$v(t) = f(u(t)), \tag{2.1}$$

N is a constant, and

$$\hat{v}(t) = Nu(t). \tag{2.2}$$

The problem is to approximate $f(u)$ by $Nu(t)$ so that

$$\varepsilon(N) = E\left[\left(v(t) - \hat{v}(t)\right)^2\right] \tag{2.3}$$

is minimized, where E denotes the expectation. The solution of this problem is given by the following theorem:

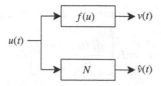

Figure 2.1. Stochastic linearization of an isolated nonlinearity.

Theorem 2.1. *If $u(t)$ is a zero-mean wide sense stationary Gaussian process and $f(u)$ is an odd, piecewise differentiable function, (2.3) is minimized by*

$$N = E\left[\left.\frac{df(u)}{du}\right|_{u=u(t)}\right]. \tag{2.4}$$

Since the proof of this theorem is simple and instructive, we provide it here, rather than in Chapter 8.

Proof. Rewriting (2.3) as

$$\varepsilon(N) = E\left[(f(u) - Nu)^2\right] \tag{2.5}$$

and differentiating with respect to N, results in the following condition of optimality:

$$\frac{d\varepsilon}{dN} = E[2(f(u) - Nu)u] = 0. \tag{2.6}$$

It is easy to verify that this is, in fact, the condition of minimality and, therefore, the minimizer of (2.3) is given by

$$N = \frac{E[f(u)u]}{E[u^2]}. \tag{2.7}$$

Taking into account that for zero-mean wss Gaussian $u(t)$ and piecewise differentiable $f(u)$,

$$E[f(u)u] = E\left[u^2\right]E\left[\left.\frac{df}{du}\right|_{u=u(t)}\right], \tag{2.8}$$

(2.4) follows immediately from the last two expressions.

$\square$

It turns out that (2.4) holds for a more general approximation of $f(u)$. To show this, let $n(t)$ be the impulse response of a causal linear system and, instead of (2.2), introduce the approximation

$$\hat{v}(t) = n(t) * u(t), \tag{2.9}$$

where $*$ denotes the convolution. The problem is to select $n(t)$ so that the functional

$$\varepsilon(n(t)) = E\left[(f(u(t)) - n(t) * u(t))^2\right] \tag{2.10}$$

is minimized.

Theorem 2.2. *Under the assumptions of Theorem 2.1, $\varepsilon(n(t))$ is minimized by*

$$n(t) = E\left[\left.\frac{df(u)}{du}\right|_{u=u(t)}\right]\delta(t), \tag{2.11}$$

where $\delta(t)$ is the δ-function.

Proof. See Section 8.1.

Thus, in this formulation as well, the minimizer of the mean square error is a *static* system with gain

$$N = E\left[\left.\frac{df(u)}{du}\right|_{u=u(t)}\right]. \tag{2.12}$$

The gain N is referred to as the *stochastic linearization* or the *quasilinear gain* of $f(u)$. Unlike the local, Jacobian, linearization of $f(u)$, that is,

$$N_J = \left.\frac{df(u)}{du}\right|_{u=u*}, \tag{2.13}$$

where u^* is an operating point, N of (2.12) is *global* in the sense that it characterizes $f(u)$ at every point with the weight defined by the statistics of $u(t)$. *This is the main utility of stochastic linearization from the point of view of the problems considered in this volume.*

Since the expectation in (2.12) is with respect to a Gaussian probability density function (pdf) defined by a single parameter – the standard deviation, σ_u, the quasilinear gain N is, in fact, a function of σ_u, that is,

$$N = N(\sigma_u). \tag{2.14}$$

With this interpretation, the quasilinear gain N can be understood as an analogue of the describing function $F(A)$ of $f(u)$, where the role of the amplitude, A, of the harmonic input

$$u(t) = A\sin\omega t$$

is played by σ_u. It is no surprise, therefore, that the accuracy of stochastic linearization is similar to that of the harmonic balance method.

As follows from (2.12), N is a linear functional of $f(u)$. This implies that if N_1 and N_2 are quasilinear gains of $f_1(u)$ and $f_2(u)$, respectively, then $N_1 + N_2$ is the quasilinear gain of $f_1(u) + f_2(u)$. Note, however, that the quasilinear gain of $\gamma f(\cdot)$, where γ is a constant, is not equal to the quasilinear gain of $f(\cdot)\gamma$: if, in a serial connection, γ precedes $f(\cdot)$ the quasilinear gain of $f(\cdot)\gamma$ is $N(\gamma\sigma_u)$; if $f(\cdot)$ precedes γ the quasilinear gain of $\gamma f(\cdot)$ is $\gamma N(\sigma_u)$. In general, of course,

$$\gamma N(\sigma_u) \neq N(\gamma\sigma_u). \tag{2.15}$$

This is why N is referred to as the quasilinear, rather than the linear, gain of $f(\cdot)$.

Examples: The stochastic linearization for typical nonlinearities of actuators and sensors is carried out below. The illustrations are provided in Table 2.1.

Saturation nonlinearity: Consider the saturation function defined by

$$f(u) = \operatorname{sat}_\alpha(u) := \begin{cases} +\alpha, & u > +\alpha, \\ u, & -\alpha \le u \le +\alpha, \\ -\alpha, & u < -\alpha, \end{cases} \tag{2.16}$$

where $\alpha > 0$. Since

$$f'(u) = \begin{cases} 1, & -\alpha < u < +\alpha, \\ 0, & u > +\alpha \text{ or } u < -\alpha, \end{cases} \tag{2.17}$$

and u is zero-mean Gaussian, it follows that

$$N = \int_{-\infty}^{+\infty} \frac{d}{du} \operatorname{sat}_\alpha(u) \frac{1}{\sqrt{2\pi}\sigma_u} \exp\left(-\frac{u^2}{2\sigma_u^2}\right) du$$

Table 2.1 *Common nonlinearities and their stochastic linearizations*

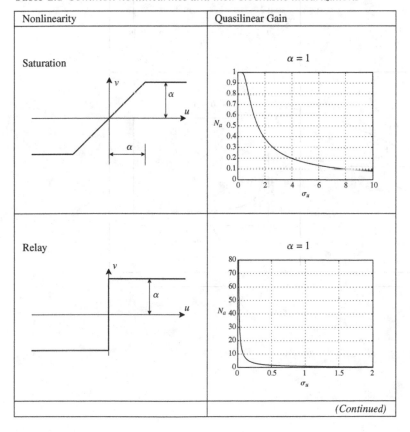

Nonlinearity	Quasilinear Gain
Saturation	
Relay	

(Continued)

Table 2.1 *(continued)*

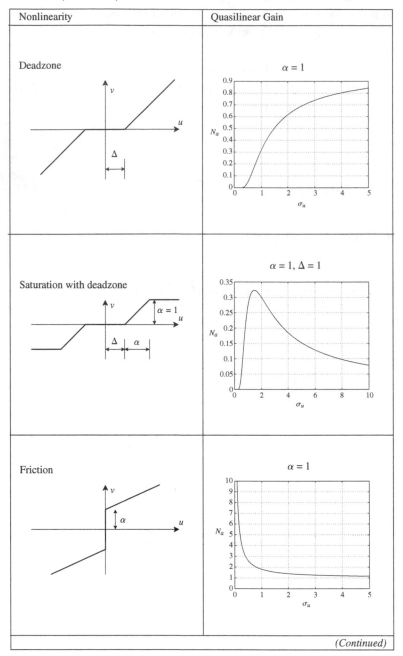

Nonlinearity	Quasilinear Gain

(Continued)

Table 2.1 *(continued)*

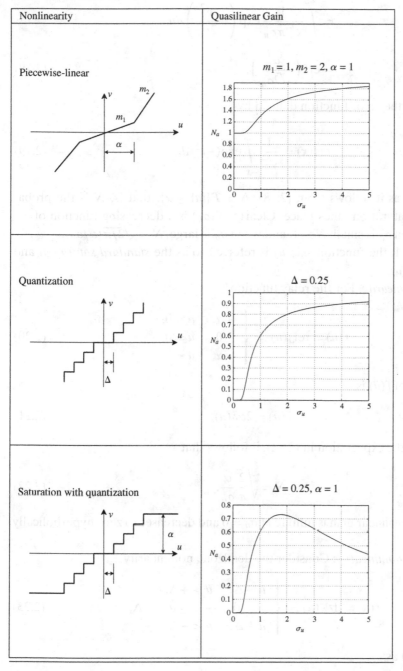

$$= \int_{-\alpha}^{+\alpha} \frac{1}{\sqrt{2\pi}\sigma_u} \exp\left(-\frac{u^2}{2\sigma_u^2}\right) du$$

$$= \mathrm{erf}\left(\frac{\alpha}{\sqrt{2}\sigma_u}\right), \tag{2.18}$$

where $\mathrm{erf}(x)$ is the error function defined by

$$\mathrm{erf}(x) = \frac{1}{\sqrt{\pi}} \int_{-x}^{+x} \exp(-t^2)\, dt. \tag{2.19}$$

Note that, as it follows from (2.18), $N = P\{|u| \leq \alpha\}$, that is, N is the probability that no saturation takes place. Clearly, $N(\sigma_u)$ is a decreasing function of σ_u. Moreover, when σ_u is small, $N \approx 1$, and when σ_u is large, $N \approx \sqrt{2/\pi}(\alpha/\sigma_u)$.

When $\alpha = 1$, the function $\mathrm{sat}_1(u)$ is referred to as the *standard saturation* and denoted as $\mathrm{sat}(u)$.

Relay nonlinearity: For the relay function,

$$f(u) = \mathrm{rel}_\alpha(u) := \begin{cases} +\alpha, & u > 0, \\ 0, & u = 0, \\ -\alpha, & u < 0, \end{cases} \tag{2.20}$$

the derivative of $f(u)$ is

$$f(u) = 2\alpha\delta(u). \tag{2.21}$$

Hence, taking the expectation in (2.12), it follows that

$$N = \sqrt{\frac{2}{\pi}} \frac{\alpha}{\sigma_u}. \tag{2.22}$$

Clearly, the quasilinear gain is infinite at $\sigma_u = 0$ and decreases to zero hyperbolically as $\sigma_u \to \infty$.

Deadzone nonlinearity: Consider the deadzone nonlinearity,

$$f(u) = \mathrm{dz}_\Delta(u) := \begin{cases} u - \Delta, & u > +\Delta, \\ 0, & -\Delta \leq u \leq +\Delta, \\ u + \Delta, & u < -\Delta. \end{cases} \tag{2.23}$$

Writing $\mathrm{dz}_\Delta(u)$ as

$$\mathrm{dz}_\Delta(u) = u - \mathrm{sat}_\Delta(u), \tag{2.24}$$

we obtain

$$N = 1 - \mathrm{erf}\left(\frac{\Delta}{\sqrt{2}\sigma_u}\right). \tag{2.25}$$

Note that $N = P\{|u| \geq \Delta\}$. Obviously, for $\sigma_u << \Delta$, $N \approx 0$, while $N \to 1$ as $\sigma_u \to \infty$.

Saturation with deadzone nonlinearity: Consider the saturated deadzone nonlinearity,

$$f(u) = \mathrm{sat}_\alpha(\mathrm{dz}_\Delta(u)) := \begin{cases} +\alpha, & u > +\alpha + \Delta, \\ u - \Delta, & +\Delta \leq u \leq +\alpha + \Delta, \\ 0, & -\Delta < u < +\Delta, \\ u + \Delta, & -\alpha - \Delta \leq u \leq -\Delta, \\ -\alpha, & u < -\alpha - \Delta. \end{cases} \tag{2.26}$$

Since

$$f'(u) = \begin{cases} 1, & \Delta < |u| < \alpha + \Delta, \\ 0, & \text{otherwise,} \end{cases} \tag{2.27}$$

it follows that

$$N = \int\limits_{-\alpha - \Delta}^{-\Delta} \frac{1}{\sqrt{2\pi}\sigma_u} \exp\left(-\frac{u^2}{2\sigma_u^2}\right) du + \int\limits_{+\Delta}^{+\alpha + \Delta} \frac{1}{\sqrt{2\pi}\sigma_u} \exp\left(-\frac{u^2}{2\sigma_u^2}\right) du$$

$$= \mathrm{erf}\left(\frac{\alpha + \Delta}{\sqrt{2}\sigma_u}\right) - \mathrm{erf}\left(\frac{\Delta}{\sqrt{2}\sigma_u}\right). \tag{2.28}$$

A characteristic feature of this nonlinearity is that N is a nonmonotonic function of σ_u: increasing for small σ_u and decreasing for large ones.

Friction nonlinearity: Consider the friction nonlinearity,

$$f(u) = \mathrm{fri}_\alpha(u) := \begin{cases} u + \alpha, & u > 0, \\ 0, & u = 0, \\ u - \alpha, & u < 0. \end{cases} \tag{2.29}$$

Since

$$\mathrm{fri}_\alpha(u) = u + \mathrm{rel}_\alpha(u), \tag{2.30}$$

it follows that

$$N = 1 + \sqrt{\frac{2}{\pi}} \frac{\alpha}{\sigma_u}. \tag{2.31}$$

Again, $N = \infty$ for $\sigma_u = 0$ and decreases to 1 hyperbolically as $\sigma_u \to \infty$.

Piecewise-linear function: For the piecewise-linear function,

$$f(u) = \mathrm{pwl}_\alpha(u) := \begin{cases} m_2 u + (m_1 - m_2)\alpha, & u > +\alpha, \\ m_1 u, & -\alpha \leq u \leq +\alpha, \\ m_2 u + (m_2 - m_1)\alpha, & u < -\alpha, \end{cases} \tag{2.32}$$

the derivative of $f(u)$ is

$$f'(u) = \begin{cases} m_1, & |u| < \alpha, \\ m_2, & |u| > \alpha. \end{cases} \tag{2.33}$$

Thus,

$$N = m_2 + (m_1 - m_2)\mathrm{erf}\left(\frac{\alpha}{\sqrt{2}\sigma_u}\right). \tag{2.34}$$

Note that $N = m_2 + (m_1 - m_2)P\{|u| \le \alpha\}$.

Quantization nonlinearity: The quantization nonlinearity is defined as

$$f(u) = \mathrm{qn}_\Delta(u) := \begin{cases} +\Delta\lfloor +u/\Delta \rfloor, & u \ge 0, \\ -\Delta\lfloor -u/\Delta \rfloor, & u < 0, \end{cases} \tag{2.35}$$

where Δ is the quantizer step size and $\lfloor u \rfloor$ denotes the largest integer less than or equal to u. Clearly,

$$f'(u) = \sum_{\substack{k=-\infty \\ k \ne 0}}^{+\infty} \Delta\delta(u - k\Delta). \tag{2.36}$$

Hence,

$$N = \frac{2\Delta}{\sqrt{2\pi\sigma_u^2}} \sum_{k=1}^{\infty} \exp\left(-\frac{\Delta^2}{2\sigma_u^2}k^2\right). \tag{2.37}$$

For $\sigma_u < \Delta$, N is akin to the deadzone and approaches 1 as $\sigma_u \to \infty$.

Saturation with quantization nonlinearity: The saturated quantization is defined as

$$f(u) = \mathrm{sat}_\alpha(\mathrm{qn}_\Delta(u)) := \begin{cases} +\alpha, & u \ge +\alpha, \\ +\Delta\lfloor +u/\Delta \rfloor, & 0 \le u < \alpha, \\ -\Delta\lfloor -u/\Delta \rfloor, & -\alpha \le u < 0, \\ -\alpha, & u < -\alpha, \end{cases} \tag{2.38}$$

where it is assumed that $\alpha = m\Delta$ for a positive integer m. The derivative of $f(u)$ is

$$f'(u) = \sum_{\substack{k=-m \\ k \ne 0}}^{+m} \Delta\delta(u - k\Delta). \tag{2.39}$$

Thus,

$$N = \frac{2\Delta}{\sqrt{2\pi\sigma_u^2}} \sum_{k=1}^{m} \exp\left(-\frac{\Delta^2}{2\sigma_u^2}k^2\right). \tag{2.40}$$

Here, again, N is nonmonotonic in σ_u.

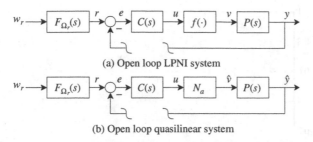

(a) Open loop LPNI system

(b) Open loop quasilinear system

Figure 2.2. Open loop LPNI system and its quasilinearization.

2.1.2 Stochastic Linearization of Direct Paths of LPNI Systems

Quasilinear gain: Consider the open loop LPNI system shown in Figure 2.2(a), where $F_{\Omega_r}(s)$, $P(s)$, and $C(s)$ are transfer functions with all poles in open left half plane (OLHP) representing the coloring filter with the 3dB bandwidth Ω_r, the plant, and the controller, respectively, $f(u)$ is the actuator nonlinearity, and w_r and r are standard white noise and the reference signal.

Since, in such a system, the steady state input to the nonlinearity is still a zero-mean wss Gaussian process, its stochastically linearized gain remains the same as in Subsection 2.1.1, and the corresponding quasilinear system is shown in Figure 2.2(b). Since the standard deviation σ_u is given by the 2-norm of the transfer function from w_r to u, that is,

$$\sigma_u = \|F_{\Omega_r}(s)C(s)\|_2 = \sqrt{\frac{1}{2\pi} \int_{-\infty}^{+\infty} |F_{\Omega_r}(j\omega)|^2 |C(j\omega)|^2 \, d\omega,}$$

the stochastically linearized gain in Figure 2.2(b) is

$$N_a(\sigma_u) = \mathcal{F}\left(\|F_{\Omega_r}(s)C(s)\|_2\right), \tag{2.41}$$

where

$$\mathcal{F}(\sigma) = \int_{-\infty}^{+\infty} \left[\frac{d}{dx}f(x)\right] \frac{1}{\sqrt{2\pi}\sigma} \exp\left(-\frac{x^2}{2\sigma^2}\right) dx. \tag{2.42}$$

Computational issues: To evaluate the standard deviation of the signal at the input of the actuator in Figure 2.2(b), one has to evaluate the 2-norm of a transfer function. A computationally convenient way to carry this out is as follows:

Let $\{A, B, C\}$ be a minimal realization of the strictly proper transfer function $G(s)$ with all poles in the open left half plane. Consider the Lyapunov equation

$$AR + RA^T + BB^T = 0. \tag{2.43}$$

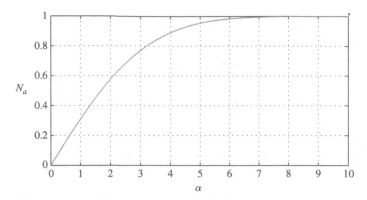

Figure 2.3. Quasilinear gain as a function of α in Example 2.1.

Then it is well known that its solution, R, is positive definite symmetric and, moreover,

$$\|G(s)\|_2 = \left[\operatorname{tr} \left(CRC^T \right) \right]^{1/2}. \tag{2.44}$$

Clearly, this provides a constructive way for calculating $\|G(s)\|_2$. In the MATLAB computational environment, the function **norm** can be used for this purpose.

Example 2.1. Consider the system of Figure 2.2(a), with

$$P(s) = \frac{10}{s(s+10)}, \; C(s) = 5, \; F_{\Omega_r}(s) = \frac{\sqrt{3}}{s^3 + 2s^2 + 2s + 1}, \; f(u) = \operatorname{sat}_\alpha(u). \tag{2.45}$$

The closed loop version of this system is a servomechanism with the reference signal, r, defined by a 3rd order Butterworth filter with bandwidth $\Omega_r = 1$ and d.c. gain selected so that $\sigma_r = 1$.

For this system, taking into account (2.18), from (2.41),

$$N_a = \operatorname{erf} \left(\frac{\alpha}{\sqrt{2} \left\| \frac{5\sqrt{3}}{s^3 + 2s^2 + 2s + 1} \right\|_2} \right). \tag{2.46}$$

The behavior of N_a as a function of α is illustrated in Figure 2.3. From this figure we conclude the following:

- For $\alpha \in (0, 2)$, $N_a(\alpha)$ is practically linear with slope 0.3.
- For $\alpha > 7$, $N_a(\alpha)$ is practically 1, that is, the effect of saturation may be ignored.

2.2 Stochastic Linearization of Closed Loop LPNI Systems

2.2.1 Notations and Assumptions

The block diagram of the LPNI feedback systems studied in this volume is shown in Figure 2.4. As in Figure 2.2, $P(s)$ and $C(s)$ are the plant and controller, $f(\cdot)$ and $g(\cdot)$ are

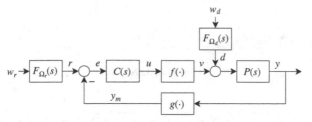

Figure 2.4. Closed loop LPNI system.

the actuator and sensor, $F_{\Omega_d}(s)$ and $F_{\Omega_r}(s)$ are coloring filters with 3dB bandwidths Ω_d and Ω_r, respectively, w_r, w_d are independent standard Gaussian white noise processes, and the scalars r, d, y, y_m, u, v, and e denote the reference, disturbance, plant output, sensor (or measured) output, control signal, actuator output, and error signal, respectively.

Let the quasilinear gains of $f(\cdot)$ and $g(\cdot)$ in isolation be denoted as N_a and N_s, respectively. Assume that the range of N_a is $\mathcal{N}_a$, the range of N_s is $\mathcal{N}_s$, and the range of $N_a N_s$ is $\mathcal{N}_{as}$. For instance, if $f(\cdot)$ and $g(\cdot)$ are standard saturation functions, then $\mathcal{N}_a = \mathcal{N}_s = \mathcal{N}_{as} = [0, 1]$. Using these notations, introduce the following:

Assumption 2.1.

(i) *$P(s)$ has all poles in the closed left half plane;*
(ii) *$C(s)$ has all poles in the closed left half plane;*
(iii) *$1 + \gamma P(s)C(s)$ has all zeros in the open left half plane for $\gamma \in \mathcal{N}_{as}$.*

This assumption, as it is shown below, is a sufficient condition for the existence of variances of various signals in the stochastically linearized versions of LPNI systems under consideration. Therefore, unless stated otherwise, it is assumed to hold throughout this volume.

Under Assumption 2.1, we discuss below stochastic linearization of the closed loop system of Figure 2.4. First, we address the case of nonlinear actuators and sensors separately and then the case of nonlinearities in both actuators and sensors simultaneously.

2.2.2 Reference Tracking with Nonlinear Actuator

Quasilinear gain: Consider the closed loop system of Figure 2.5(a), where all functional blocks and signals remain the same as in Figure 2.4. The goal is to obtain its quasilinear approximation shown in Figure 2.5(b).

The situation here is different from that of Figure 2.2 in two respects. First, the signal $u(t)$ at the input of the nonlinearity is no longer Gaussian. Second, the signals $u(t)$ and $\hat{u}(t)$ are not the same. Therefore, the quasilinear gain (2.12) is no longer optimal. Nevertheless, proceeding formally, we view the system of Figure 2.5(b) with

$$N_a = E\left[\frac{df(\hat{u})}{d\hat{u}}\bigg|_{\hat{u}=\hat{u}(t)} \right] \tag{2.47}$$

as the stochastic linearization of Figure 2.5(a).

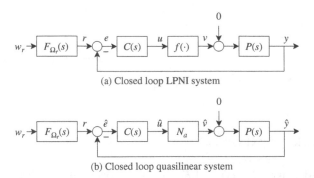

(a) Closed loop LPNI system

(b) Closed loop quasilinear system

Figure 2.5. Reference tracking closed loop LPNI system with nonlinear actuator and its quasilinearization.

Although the accuracy of this approximation is discussed in Section 2.3, we note here that the first of the above obstacles to optimality is alleviated by the fact that, if the plant is low-pass filtering, the signal $u(t)$ is close to Gaussian, even if $v(t)$ is not. The second obstacle, however, is unavoidable. Nevertheless, as shown in Section 2.3 and numerous previous studies, stochastic linearization of closed loop systems results in accuracy well within 10%, as far as the difference between the standard deviations of the outputs, σ_y and $\sigma_{\hat{y}}$, is concerned.

Since the standard deviation of $\hat{u}$ is

$$\sigma_{\hat{u}} = \left\| \frac{F_{\Omega_r}(s)C(s)}{1+P(s)N_aC(s)} \right\|_2 ,$$

it follows from (2.12) that the quasilinear gain of Figure 2.5(b) is defined by

$$N_a = \mathcal{F}\left(\left\| \frac{F_{\Omega_r}(s)C(s)}{1+P(s)N_aC(s)} \right\|_2 \right), \tag{2.48}$$

where

$$\mathcal{F}(\sigma) = \int_{-\infty}^{\infty} \left[\frac{d}{dx} f(x) \right] \frac{1}{\sqrt{2\pi}\sigma} \exp\left(-\frac{x^2}{2\sigma^2} \right) dx. \tag{2.49}$$

Thus, N_a is a root of the equation

$$x - \mathcal{F}\left(\left\| \frac{F_{\Omega_r}(s)C(s)}{1+xP(s)C(s)} \right\|_2 \right) = 0, \tag{2.50}$$

which is referred to as the *reference tracking quasilinear gain equation*. This equation is used in Chapters 3 and 5 for analysis and design of LPNI systems from the point of view of reference tracking.

Note that, as indicated in Chapter 1, the quasilinear system of Figure 2.5(b) with N_a defined by (2.50) is a *global* approximation of the LPNI system of Figure 2.5(a). In comparison with the *local*, Jacobian, linearization, the price to pay is that the linearized gain depends not only on the nonlinearity but on all other functional blocks of the system, that is, $F_{\Omega_r}(s)$, $C(s)$, $P(s)$, as well as on the exogenous signal w_r.

Equation (2.50) defines N_a as an implicit function of $P(s)$, $C(s)$, and $F_{\Omega_r}(s)$. A sufficient condition for the existence of its solution is given by the following theorem:

Theorem 2.3. *Under Assumption 2.1, the transcendental equation (2.50) has a solution.*

Proof. See Section 8.1.

Computational issues: The solution of (2.50) amounts to finding intersections of the linear function x with the nonlinear function

$$\mathcal{F}\left(\left\|\frac{F_{\Omega_r}(s)C(s)}{1+xP(s)C(s)}\right\|_2\right).$$

Although various algorithms can be used for this purpose, we found it convenient to use a bisection approach. To accomplish this, assume that $N_a(\sigma_{\hat{u}})$ is bounded for $\sigma_{\hat{u}} \in [0,\infty)$ and the solution of (2.50) is unique. Then, to find the solution, x^*, when $\mathcal{F}(x) > x$ for $x < x^*$, we formulate the following:

Bisection Algorithm 2.1. Given a desired accuracy $\epsilon > 0$:

(a) Set the variables

$$N^- = \min_{\sigma_{\hat{u}} \in [0,\infty)} N_a(\sigma_{\hat{u}})$$

$$N^+ = \max_{\sigma_{\hat{u}} \in [0,\infty)} N_a(\sigma_{\hat{u}})$$

(b) If $|N^+ - N^-| < \epsilon$, go to step (f)
(c) Let $N = (N^- + N^+)/2$
(d) Evaluate the left-hand side of (2.50) with $x = N$, call it δ
(e) If $\delta < 0$, let $N^- = N$, else let $N^+ = N$, and go to step (c)
(f) N is an ϵ-precise solution of (2.50).

To find the solution, x^*, when $\mathcal{F}(x) < x$ for $x < x^*$, we use the same algorithm but with the step (e) substituted by the following step:

(e') If $\delta < 0$, let $N^+ = N$, else let $N^- = N$, and go to step (2.1)

If $N_a(\sigma_{\hat{u}})$, $\sigma_{\hat{u}} \in [0,\infty)$ is unbounded and/or (2.50) has multiple solutions, one can attempt to find them by repeated applications of Bisection Algorithm 2.1 choosing different initial conditions.

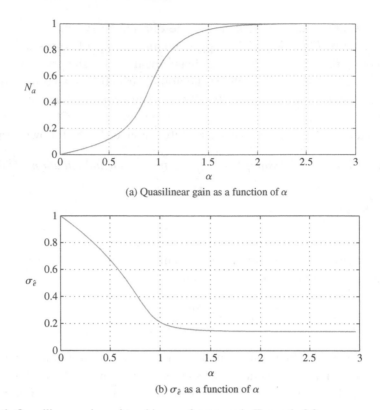

(a) Quasilinear gain as a function of α

(b) $\sigma_{\hat{e}}$ as a function of α

Figure 2.6. Quasilinear gain and tracking performance in Example 2.2.

Example 2.2. Consider the system of Figure 2.5(a) with all elements defined as in (2.45). For this system, (2.50) becomes

$$x - \text{erf}\left(\frac{\alpha}{\sqrt{2}\left\|\frac{5\sqrt{3}s(s+10)}{(s^3+2s^2+2s+1)(s^2+10s+50x)}\right\|_2}\right) = 0. \tag{2.51}$$

Using Bisection Algorithm 2.1, we solve this equation for $\alpha \in [0,3]$. Then, using the Lyapunov equation (2.43), we calculate the standard deviation of the tracking error $\sigma_{\hat{e}}$. The results are shown in Figure 2.6. From this figure we conclude (compare with the open-loop case of Figure 2.3) the following:

- For $\alpha \in (0,0.5)$, $N_a(\alpha)$ is practically linear with slope 0.2. With these α's, the tracking error, $\sigma_{\hat{e}}$, is decreasing, also in a practically linear manner, with slope (-0.7).
- For $\alpha > 2$, $N_a(\alpha)$ is practically 1, and the tracking error is the same as in the corresponding linear system (i.e., when the saturation is ignored): $\sigma_{\hat{e}} = 0.142$.

The graphs of Figure 2.6 characterize the tracking performance of the quasilinear system of Figure 2.5(b). To illustrate how this performance compares with that of the original nonlinear system of Figure 2.5(a), Figure 2.7 shows traces of $r(t)$, $y(t)$,

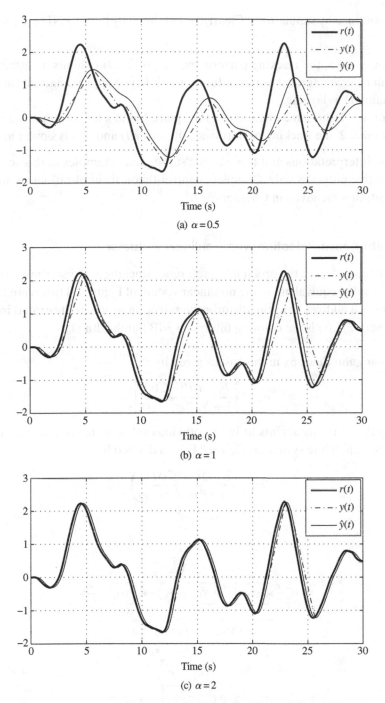

(a) $\alpha = 0.5$

(b) $\alpha = 1$

(c) $\alpha = 2$

Figure 2.7. Illustration of reference tracking in Example 2.2 in time domain.

and $\hat{y}(t)$ obtained by simulations. Clearly, as exhibited by both LPNI and quasilinear systems,

- For $\alpha = 0.5$, the tracking performance is poor: $y(t)$ exhibits an effect, which can be characterized as rate saturation, while $\hat{y}(t)$ approximates it (in a linear manner) as lagging.
- For $\alpha = 1$, the same phenomena take place but in a less pronounced manner.
- For $\alpha = 2$, the tracking performance of both $y(t)$ and $\hat{y}(t)$ is similar and good.

These interpretations and those of the subsequent examples in this section are for illustrative purposes only. Complete control-theoretic implications of nonlinear instrumentation are given in Chapters 3–7.

2.2.3 Disturbance Rejection with Nonlinear Actuator

While in linear systems, the problems of reference tracking and disturbance rejection can be viewed as equivalent, in the nonlinear system of Figure 2.4 this is not the case. Therefore, we address here the disturbance rejection problem illustrated in Figure 2.8(a), where $F_{\Omega_d}(s)$ is the coloring filter with 3dB bandwidth Ω_d.

Quasilinear gain: Since $\sigma_{\hat{u}}$ in this case is given by

$$\sigma_{\hat{u}} = \left\| \frac{F_{\Omega_d}(s)P(s)C(s)}{1 + N_a P(s) C(s)} \right\|_2,$$

following the same arguments as in the previous subsection, we conclude that the quasilinear gain in the system of Figure 2.8(b) is defined by

$$x - \mathcal{F}\left(\left\| \frac{F_{\Omega_d}(s)P(s)C(s)}{1 + x P(s) C(s)} \right\|_2 \right) = 0. \tag{2.52}$$

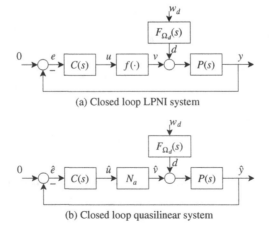

(a) Closed loop LPNI system

(b) Closed loop quasilinear system

Figure 2.8. Disturbance rejection closed loop LPNI system with nonlinear actuator and its quasilinear approximation.

This is referred to as the *disturbance rejection quasilinear gain equation*. The existence of its solution can be guaranteed by a theorem similar to Theorem 2.3, and the solution can be found using Bisection Algorithm 2.1. We use this equation in Chapters 4 and 6 for analysis and design of LPNI systems from the point of view of disturbance rejection.

Example 2.3. Consider the system of Figure 2.8(a) with the same parameters as in (2.45), that is,

$$P(s) = \frac{10}{s(s+10)}, \quad C(s) = 5, \quad F_{\Omega_d}(s) = \frac{\sqrt{3}}{s^3 + 2s^2 + 2s + 1}, \quad f(u) = \mathrm{sat}_\alpha(u). \quad (2.53)$$

For this system, (2.52) becomes

$$x - \mathrm{erf}\left(\frac{\alpha}{\sqrt{2}\left\| \frac{50\sqrt{3}}{(s^3 + 2s^2 + 2s + 1)(s^2 + 10s + 50x)} \right\|_2} \right) = 0. \quad (2.54)$$

Using Bisection Algorithm 2.1, we solve this equation for $\alpha \in [0,3]$ and then calculate the disturbance rejection performance in terms of the output standard deviation $\sigma_{\hat{y}}$. The results are shown in Figure 2.9. From this figure we conclude the following:

- The quasilinear gains of the disturbance rejection problem and reference tracking problem are quite different, even when all functional blocks remain the same (compare Figures 2.6(a) and 2.9(a)). Namely, in this example, N_a for the disturbance rejection problem is smaller than N_a for the reference tracking problem.
- For $\alpha \in (0,0.5]$, $N_a(\alpha)$ is practically linear with the slope 0.025. With these α's, disturbance rejection is poor: $\sigma_{\hat{y}}$ ranges from 50 to 2 (recall that $\sigma_d = 1$).
- For $\alpha > 2.5$, $N_a(\alpha)$ is practically 1 and $\sigma_{\hat{y}}$ takes values close to those when the saturation is ignored: $\sigma_{\hat{y}} = 0.2$.

The traces of $d(t)$, $y(t)$, and $\hat{y}(t)$ obtained by simulations are shown in Figure 2.10. Clearly, the disturbance rejection improves with an increasing α, and $\hat{y}(t)$ approximates $y(t)$ sufficiently well.

2.2.4 Reference Tracking and Disturbance Rejection with Nonlinear Sensor

The quasilinear gain of the system of Figure 2.11 is defined by

$$N_s = E\left[\frac{dg(\hat{y})}{d\hat{y}}\bigg|_{\hat{y}=\hat{y}(t)} \right]. \quad (2.55)$$

Similar to the previous two subsections the following equations for N_s can be derived:
Reference tracking quasilinear gain equation:

$$x - \mathcal{G}\left(\left\| \frac{F_{\Omega_r}(s)C(s)P(s)}{1 + xP(s)C(s)} \right\|_2 \right) = 0. \quad (2.56)$$

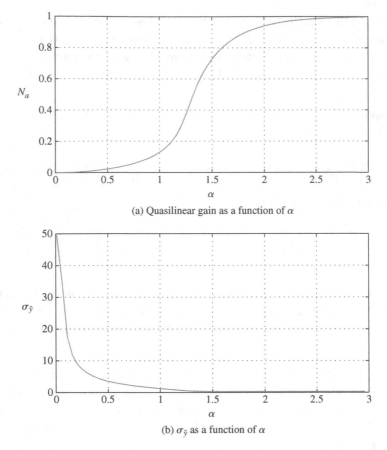

(a) Quasilinear gain as a function of α

(b) $\sigma_{\hat{y}}$ as a function of α

Figure 2.9. Quasilinear gain and disturbance rejection performance in Example 2.3.

Note that this equation has the same structure as the disturbance rejection quasilinear gain equation with nonlinear actuator, where

$$\mathcal{G}(\sigma) = \int_{-\infty}^{+\infty} \left[\frac{d}{dx} g(x) \right] \frac{1}{\sqrt{2\pi}\sigma} \exp\left(-\frac{x^2}{2\sigma^2} \right) dx. \tag{2.57}$$

Disturbance rejection quasilinear gain equation:

$$x - \mathcal{G}\left(\left\| \frac{F_{\Omega_d}(s)P(s)}{1 + xP(s)C(s)} \right\|_2 \right) = 0. \tag{2.58}$$

Solutions of these equations can be found using Bisection Algorithm 2.1 of Subsection 2.2.2.

Example 2.4. Consider the system of Figure 2.11(b) with all linear blocks as in (2.45), that is,

$$P(s) = \frac{10}{s(s+10)}, \quad C(s) = 5, \quad F_{\Omega_d}(s) = \frac{\sqrt{3}}{s^3 + 2s^2 + 2s + 1}, \tag{2.59}$$

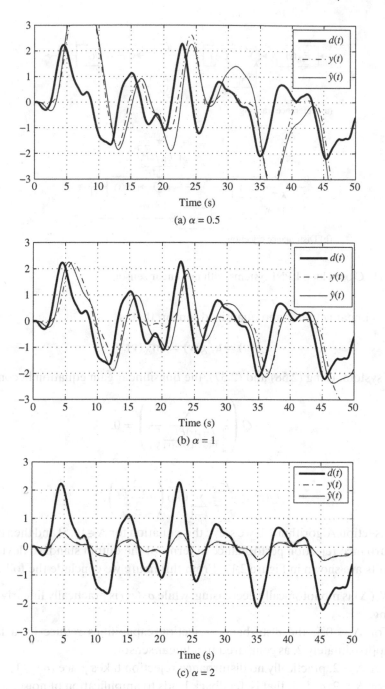

Figure 2.10. Illustration of disturbance rejection in Example 2.3 in time domain.

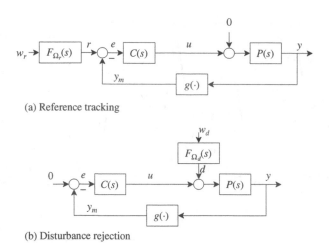

(a) Reference tracking

(b) Disturbance rejection

Figure 2.11. Closed loop LPNI systems with nonlinear sensor.

but with a linear actuator and quantized sensor, that is,

$$f(u) = u, \ g(y) = \mathrm{qn}_\Delta(y). \tag{2.60}$$

For this system, using (2.58) and (2.37), the quasilinear gain equation becomes

$$x - Q\left(\frac{\Delta}{\left\| \frac{F_{\Omega_d}(s)P(s)}{1+xP(s)C(s)} \right\|_2} \right) = 0, \tag{2.61}$$

where

$$Q(z) = \frac{\sqrt{2}z}{\sqrt{\pi}} \sum_{k=1}^{\infty} \exp\left(-\frac{z^2}{2} k^2 \right). \tag{2.62}$$

Using Bisection Algorithm 2.1, we solve this equation for $\Delta \in [0,5]$ and then calculate the disturbance rejection performance in terms of the output standard deviation $\sigma_{\hat{y}}$. The results are shown in Figure 2.12. From this figure we conclude the following:

- $N_s(\Delta)$ is monotonically decreasing, while $\sigma_{\hat{y}}(\Delta)$ is practically linearly increasing.
- For $\Delta = 0.5$, the disturbance rejection capabilities reduce by a factor of approximately 2, as compared to a linear sensor.
- For $\Delta = 2$, practically no disturbance rejection takes place ($\sigma_{\hat{y}} \approx 1$).
- For $\Delta > 2$, $\sigma_{\hat{y}} > 1$, that is, feedback leads to amplification of noise.

Time traces of $r(t)$ and $y(t)$, $\hat{y}(t)$ are shown in Figure 2.13.

2.2.5 Closed Loop LPNI Systems with Nonlinear Actuators and Sensors

Reference tracking: The block diagram of such a system for reference tracking and its quasilinear approximation are shown in Figure 2.14, where the two quasilinear

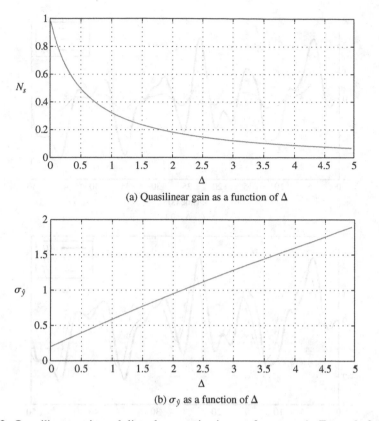

(a) Quasilinear gain as a function of Δ

(b) $\sigma_{\hat{y}}$ as a function of Δ

Figure 2.12. Quasilinear gain and disturbance rejection performance in Example 2.4.

gains, N_a and N_s, are defined by

$$N_a = E\left[\left.\frac{d}{d\hat{u}}f(\hat{u})\right|_{\hat{u}=\hat{u}(t)}\right], \tag{2.63}$$

$$N_s = E\left[\left.\frac{d}{d\hat{y}}g(\hat{y})\right|_{\hat{y}=\hat{y}(t)}\right]. \tag{2.64}$$

Following the same procedure as above, we obtain

$$N_a = \mathcal{F}\left(\left\|\frac{F_{\Omega_r}(s)C(s)}{1+P(s)N_sC(s)N_a}\right\|_2\right), \tag{2.65}$$

$$N_s = \mathcal{G}\left(\left\|\frac{F_{\Omega_r}(s)C(s)N_aP(s)}{1+P(s)N_sC(s)N_a}\right\|_2\right), \tag{2.66}$$

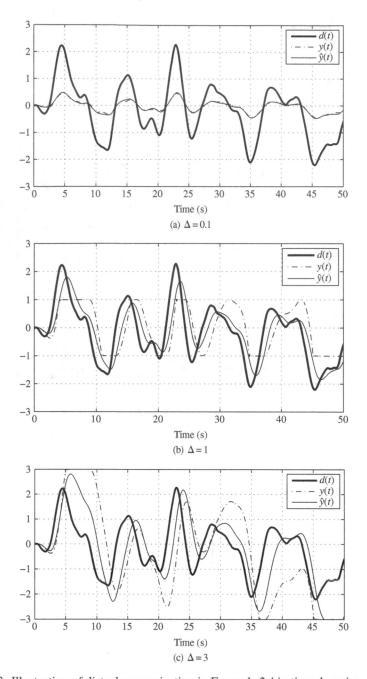

Figure 2.13. Illustration of disturbance rejection in Example 2.4 in time domain.

where

$$\mathcal{F}(\sigma) = \int\limits_{-\infty}^{\infty} \left[\frac{d}{dx} f(x) \right] \frac{1}{\sqrt{2\pi}\sigma} \exp\left(-\frac{x^2}{2\sigma^2} \right) dx, \qquad (2.67)$$

$$\mathcal{G}(\sigma) = \int\limits_{-\infty}^{\infty} \left[\frac{d}{dx} g(x) \right] \frac{1}{\sqrt{2\pi}\sigma} \exp\left(-\frac{x^2}{2\sigma^2} \right) dx. \qquad (2.68)$$

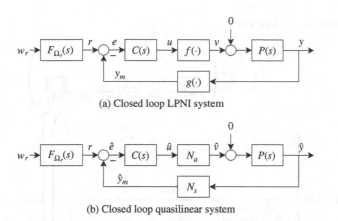

(a) Closed loop LPNI system

(b) Closed loop quasilinear system

Figure 2.14. Reference tracking closed loop LPNI system with nonlinear instrumentation and its quasilinear approximation.

Accordingly, the task of finding the quasilinear gains N_a and N_s is a two-variable root-finding problem for equations (2.65) and (2.66). Alternatively, an elimination procedure can be used to formulate an equivalent single-variable problem. Specifically, multiplying (2.65) and (2.66) results in

$$N_{as} = \mathcal{F}\left(\left\|\frac{F_{\Omega_r}(s)C(s)}{1 + N_{as}P(s)C(s)}\right\|_2\right)\mathcal{G}\left(\left\|\frac{\mathcal{F}\left(\left\|\frac{F_{\Omega_r}(s)C(s)}{1 + N_{as}P(s)C(s)}\right\|_2\right)F_{\Omega_r}(s)P(s)C(s)}{1 + N_{as}P(s)C(s)}\right\|_2\right),$$

(2.69)

where $N_{as} = N_a N_s$. Clearly, N_{as} is a root of the equation

$$x - \mathcal{F}\left(\left\|\frac{F_{\Omega_r}(s)C(s)}{1 + xP(s)C(s)}\right\|_2\right)\mathcal{G}\left(\left\|\frac{\mathcal{F}\left(\left\|\frac{F_{\Omega_r}(s)C(s)}{1 + xP(s)C(s)}\right\|_2\right)F_{\Omega_r}(s)P(s)C(s)}{1 + xP(s)C(s)}\right\|_2\right) = 0.$$

(2.70)

With N_{as} known, N_a is determined from (2.65). In turn, with N_{as} and N_a known, (2.66) determines N_s. The computational issues for solving (2.70) remain the same as in the previous subsection.

Example 2.5. Consider the LPNI system of Figure 2.14(a), with all elements as in Example 2.2, that is,

$$P(s) = \frac{10}{s(s+10)}, \quad C(s) = 5, \quad F_{\Omega_r}(s) = \frac{\sqrt{3}}{s^3 + 2s^2 + 2s + 1}, \quad f(u) = \text{sat}_\alpha(u) \quad (2.71)$$

and with

$$g(y) = \text{dz}_\Delta(y). \quad (2.72)$$

For this system, (2.70) becomes

$$x - \mathrm{erf}\left(\frac{\alpha}{\sqrt{2}\left\|\frac{5\sqrt{3}s(s+10)}{(s^3+2s^2+2s+1)(s^2+10s+50x)}\right\|_2}\right) \times$$

$$\left[1 - \mathrm{erf}\left(\frac{\Delta}{\sqrt{2}\left\|\frac{50\sqrt{3}\mathrm{erf}\left(\frac{\alpha}{\sqrt{2}\left\|\frac{5\sqrt{3}s(s+10)}{(s^3+2s^2+2s+1)(s^2+10s+50x)}\right\|_2}\right)}{(s^3+2s^2+2s+1)(s^2+10s+50x)}\right\|_2}\right)\right] = 0. \qquad (2.73)$$

Using Bisection Algorithm 2.1, it is possible to evaluate N_a and N_s as functions of α and Δ. Then, we evaluate the variance of the tracking error, $\hat{e}_{tr}(t)$, defined as

$$\hat{e}_{tr}(t) = r(t) - \hat{y}(t). \qquad (2.74)$$

Figure 2.15 illustrates N_a, N_s, and $\sigma_{\hat{e}_{tr}}$ as functions of α for $\Delta = 1$. Similarly, Figure 2.16 illustrates N_a, N_s, and $\sigma_{\hat{e}_{tr}}$ as functions of Δ for $\alpha = 1$. Time traces of $r(t)$, $y(t)$, and $\hat{y}(t)$ are shown in Figure 2.17. From these figures, we conclude the following:

- The quasilinear gains N_a and N_s are coupled. In other words, a change in actuator authority α changes not only N_a but also N_s. Similarly, changing the sensor deadzone changes both N_s and N_a.
- N_s is monotonically increasing as a function of α; N_a is monotonically decreasing as a function of Δ.
- The quality of reference tracking is nonmonotonic with respect to α (see Figure 2.15(c)).
- The quality of reference tracking may degrade dramatically if the sensor has a deadzone (compare Figures 2.6(b) and 2.15(c)).

Disturbance rejection: For the case of the disturbance rejection problem of Figure 2.18, the equations for N_a and N_s become

$$N_a = \mathcal{F}\left(\left\|\frac{F_{\Omega_d}(s)P(s)N_sC(s)}{1 + P(s)N_sC(s)N_a}\right\|_2\right), \qquad (2.75)$$

$$N_s = \mathcal{G}\left(\left\|\frac{F_{\Omega_d}(s)P(s)}{1 + P(s)N_sC(s)N_a}\right\|_2\right). \qquad (2.76)$$

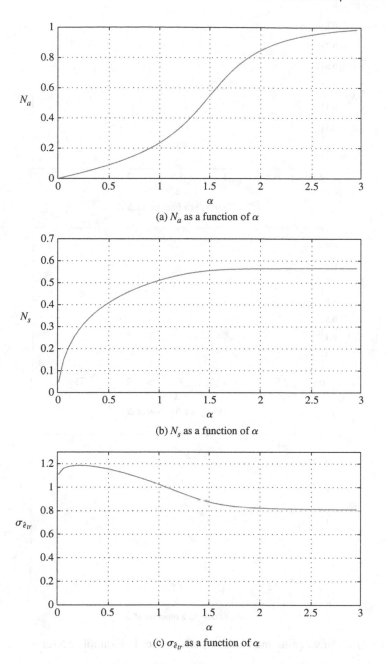

(a) N_a as a function of α

(b) N_s as a function of α

(c) $\sigma_{\hat{e}_{tr}}$ as a function of α

Figure 2.15. Quasilinear gains and tracking performance in Example 2.5 for $\Delta = 1$.

Thus, using an elimination procedure, $N_a N_s$ can be obtained from the equation

$$x - \mathcal{F}\left(\left\| \frac{\mathcal{G}\left(\left\| \frac{F_{\Omega_d}(s)P(s)}{1 + xP(s)C(s)} \right\|_2 \right) F_{\Omega_d}(s)P(s)C(s)}{1 + xP(s)C(s)} \right\|_2 \right) \mathcal{G}\left(\left\| \frac{F_{\Omega_d}(s)P(s)}{1 + xP(s)C(s)} \right\|_2 \right) = 0.$$

(2.77)

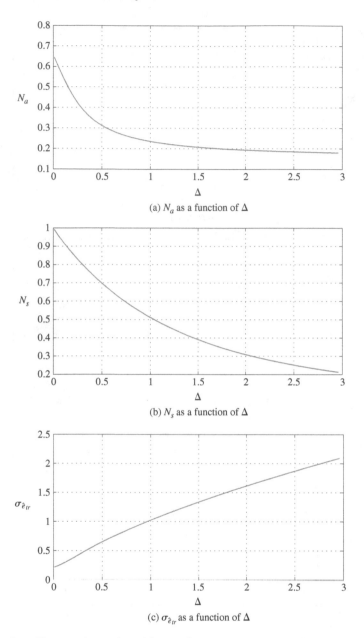

(a) N_a as a function of Δ

(b) N_s as a function of Δ

(c) $\sigma_{\hat{e}_{tr}}$ as a function of Δ

Figure 2.16. Quasilinear gains and tracking performance in Example 2.5 for $\alpha = 1$.

Equations (2.70) and (2.77) are used in subsequent chapters for analysis and design of LPNI systems with nonlinearities in both sensors and actuators.

2.2.6 Multiple Solutions of Quasilinear Gain Equations

In all examples considered above, the equations for the quasilinear gain had a unique solution. In general, this may not be the case, and in some cases multiple solutions

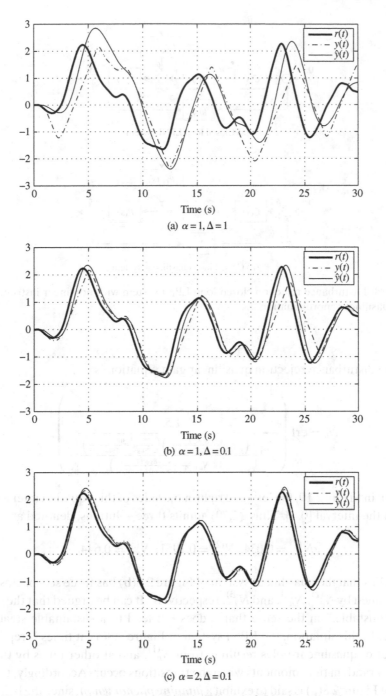

Figure 2.17. Illustration of reference tracking in Example 2.5 in time domain.

may exist. For instance, consider the LPNI system of Figure 2.8(a) with

$$P(s) = \frac{s^2 + 8s + 17}{s^3 + 4.2s^2 + 0.81s + 0.04}, \ C(s) = 1, \ f(u) = \text{sat}_{1.5}(u), \qquad (2.78)$$

$$F_{\Omega_d}(s) = \frac{0.3062}{s^3 + s^2 + 0.5s + 0.125}.$$

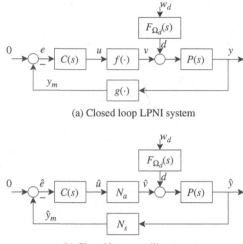

(a) Closed loop LPNI system

(b) Closed loop quasilinear system

Figure 2.18. Disturbance rejection closed loop LPNI system with nonlinear instrumentation and its quasilinear approximation.

Here, the disturbance rejection quasilinear gain equation is

$$N_a = \text{erf}\left(\frac{1.5}{\sqrt{2} \left\| \frac{\left(\frac{s^2+8s+17}{s^3+4.2s^2+0.81s+0.04} \right)\left(\frac{0.3062}{s^3+s^2+0.5s+0.125} \right)}{1+N_a\left(\frac{s^2+8s+17}{s^3+4.2s^2+0.81s+0.04} \right)} \right\|_2} \right). \tag{2.79}$$

As shown in Figure 2.19, the left- and right-hand sides of this equation intersect three times on the interval $[0,1]$. Thus, (2.79) admits three solutions, denoted as

$$N_a^{(1)} = 0.034, \ N_a^{(2)} = 0.1951, \ N_a^{(3)} = 0.644. \tag{2.80}$$

This implies that the quasilinear system of Figure 2.8(b) has three sets of closed loop poles, defined by $N_a^{(1)}$, $N_a^{(2)}$, and $N_a^{(3)}$, respectively. It can be argued that the solution $N_a^{(2)}$ is "unstable," in the sense that it does not lead to a sustainable steady state behavior. In this situation, the LPNI system of Figure 2.8(a) at times is represented by the set of quasilinear poles resulting from $N_a^{(1)}$, and at other times by that from $N_a^{(3)}$, with random time moments when the transitions occur. Accordingly, the LPNI system of Figure 2.8(a) is said to exhibit a *jumping phenomenon*, since the behavior of the system "jumps" between the possible regimes. For the system defined by (2.78), this is illustrated in Figure 2.20.

Although the average residence time in each set of the poles can be evaluated using the large deviations theory, it is not pursued here since, from the control point of view, the LPNI behavior under multiple N_a's can be characterized by the set of closed loop poles, which results in the largest $\sigma_{\hat{y}}$, thereby providing the lower bound

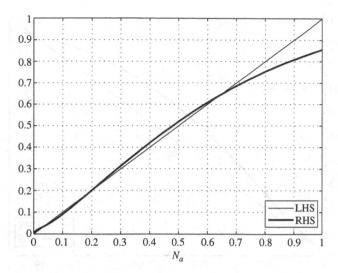

Figure 2.19. Multiple intersections of the left- and right-hand sides of (2.79) with $\alpha = 1.5$.

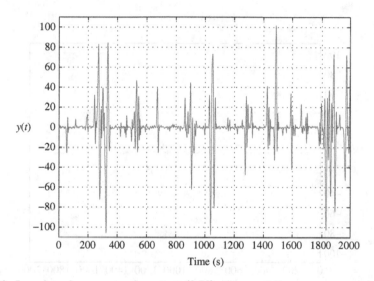

Figure 2.20. Jumping phenomenon in system (2.78) with $\alpha = 1.5$.

of the LPNI system performance. Thus, for the example of (2.78), the standard deviation of the output is not worse than $\sigma_{\hat{y}} = 35.24$ (which corresponds to $N_a^{(1)}$).

As mentioned, multiple solutions are rare in stochastic linearization and some effort is required to find a system that exhibits such behavior. Indeed, by slightly modifying (2.78) so that $\alpha = 2$, we find that the left- and right-hand sides of (2.79) exhibits a unique intersection, as shown in Figure 2.21, and the output does not exhibit the jumping behavior, as shown in Figure 2.22.

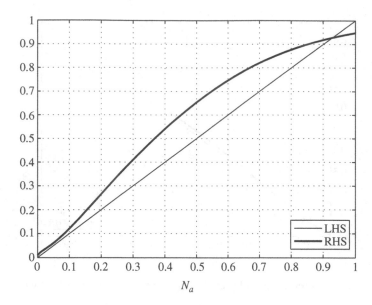

Figure 2.21. Unique intersection of the left- and right-hand sides of (2.79) with $\alpha = 2$.

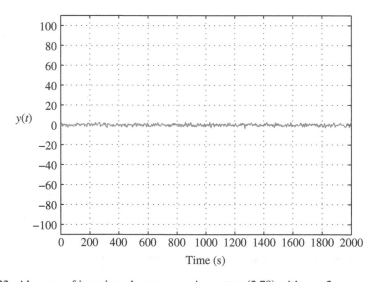

Figure 2.22. Absence of jumping phenomenon in system (2.78) with $\alpha = 2$.

2.2.7 Stochastic Linearization of State Space Equations

Stochastic linearization is readily applicable to problems formulated in state space form. Indeed, consider the LPNI system of Figure 2.4 and assume that $\{A_p, B_p, C_p\}$, $\{A_c, B_c, C_c\}$, $\{A_r, B_r, C_r\}$, and $\{A_d, B_d, C_d\}$ are minimal realizations of $P(s)$, $C(s)$, $F_{\Omega_r}(s)$, and $F_{\Omega_d}(s)$, respectively. Then, the closed loop LPNI system can be represented as

$$
\begin{bmatrix} \dot{x}_p \\ \dot{x}_c \\ \dot{x}_r \\ \dot{x}_d \end{bmatrix} = \begin{bmatrix} A_p x_p + B_p f(C_c x_c) + B_p C_d x_d \\ A_c x_c - B_c g(C_p x_p) + B_c C_r x_r \\ A_r x_r + B_r w_r \\ A_d x_d + B_d x_d \end{bmatrix},
$$

$$
\begin{bmatrix} u \\ y \end{bmatrix} = \begin{bmatrix} C_c x_c \\ C_p x_p \end{bmatrix},
$$
(2.81)

where x_p, x_c, x_r, and x_d are the states of the plant, controller, reference coloring filter, and disturbance coloring filter, respectively. Applying stochastic linearization to (2.81) results in the following quasilinear system:

$$
\begin{bmatrix} \dot{\hat{x}}_p \\ \dot{\hat{x}}_c \\ \dot{\hat{x}}_r \\ \dot{\hat{x}}_d \end{bmatrix} = \begin{bmatrix} A_p & B_p N_a C_c & 0 & B_p C_d \\ -B_c N_s C_p & A_c & B_c C_r & 0 \\ 0 & 0 & A_r & 0 \\ 0 & 0 & 0 & A_d \end{bmatrix} \begin{bmatrix} \hat{x}_p \\ \hat{x}_c \\ \hat{x}_r \\ \hat{x}_d \end{bmatrix} + \begin{bmatrix} 0 & 0 \\ 0 & 0 \\ B_r & 0 \\ 0 & B_d \end{bmatrix} \begin{bmatrix} w_r \\ w_d \end{bmatrix},
$$

$$
\begin{bmatrix} \hat{u} \\ \hat{y} \end{bmatrix} = \begin{bmatrix} 0 & C_c & 0 & 0 \\ C_p & 0 & 0 & 0 \end{bmatrix} \begin{bmatrix} \hat{x}_p \\ \hat{x}_c \\ \hat{x}_r \\ \hat{x}_d \end{bmatrix},
$$
(2.82)

where

$$
N_a = \mathcal{F}(\sigma_{\hat{u}}),
$$
(2.83)

$$
N_s = \mathcal{G}(\sigma_{\hat{y}}),
$$
(2.84)

and $\mathcal{F}(\cdot)$ and $\mathcal{G}(\cdot)$ are given in (2.67) and (2.68). Defining

$$
A = \begin{bmatrix} A_p & B_p N_a C_c & 0 & B_p C_d \\ -B_c N_s C_p & A_c & B_c C_r & 0 \\ 0 & 0 & A_r & 0 \\ 0 & 0 & 0 & A_d \end{bmatrix},
$$

$$
B = \begin{bmatrix} 0 & 0 \\ 0 & 0 \\ B_r & 0 \\ 0 & B_d \end{bmatrix},
$$
(2.85)

$$
C = \begin{bmatrix} 0 & C_c & 0 & 0 \\ C_p & 0 & 0 & 0 \end{bmatrix},
$$

it follows that the variances of $\hat{u}$ and $\hat{y}$ can be calculated using $CR(N_a, N_s)C^T$, where $R(N_a, N_s)$ is a positive definite solution of the Lyapunov equation

$$
AR(N_a, N_s) + R(N_a, N_s)A^T + BB^T = 0.
$$
(2.86)

Specifically, with

$$C_1 = \begin{bmatrix} 0 & C_c & 0 & 0 \end{bmatrix} \tag{2.87}$$

and

$$C_2 = \begin{bmatrix} C_p & 0 & 0 & 0 \end{bmatrix}, \tag{2.88}$$

it is easy to see that

$$\sigma_{\hat{u}}^2 = C_1 R(N_a, N_s) C_1^T \tag{2.89}$$

and

$$\sigma_{\hat{y}}^2 = C_2 R(N_a, N_s) C_2^T. \tag{2.90}$$

Thus, the quasilinear gain equations (2.83) and (2.84) become

$$N_a = \mathcal{F}\left(\sqrt{C_1 R(N_a, N_s) C_1^T}\right) \tag{2.91}$$

and

$$N_s = \mathcal{G}\left(\sqrt{C_2 R(N_a, N_s) C_2^T}\right). \tag{2.92}$$

Under Assumption 2.1, $R(N_a, N_s)$ is defined for every pair $(N_a, N_s) \in \mathcal{N}_a \times \mathcal{N}_s$, and (2.91), (2.92) can be solved using Bisection Algorithm 2.1.

The general equations (2.85)–(2.92) can be readily specialized to particular cases of reference tracking and disturbance rejection problems discussed in Subsections 2.2.2–2.2.5. Indeed, consider the reference tracking LPNI system with nonlinear actuator of Figure 2.5. For this case, $N_s = 1$ and $F_{\Omega_d}(s) = 0$. Thus, the matrices of (2.85) reduce to

$$A_a^r = \begin{bmatrix} A_p & B_p N_a C_c & 0 \\ -B_c C_p & A_c & B_c C_r \\ 0 & 0 & A_r \end{bmatrix},$$

$$B_a^r = \begin{bmatrix} 0 \\ 0 \\ B_r \end{bmatrix}, \tag{2.93}$$

$$C_a^r = \begin{bmatrix} 0 & C_c & 0 \\ C_p & 0 & 0 \end{bmatrix}.$$

Moreover, the *reference tracking quasilinear gain equation* becomes

$$N_a = \mathcal{F}\left(\sqrt{C_{1a}^r R_a^r(N_a) C_{1a}^{rT}}\right), \tag{2.94}$$

where

$$C_{1a}^r = \begin{bmatrix} 0 & C_c & 0 \end{bmatrix} \tag{2.95}$$

and $R_a^r(N_a)$ is a positive definite solution of

$$A_a^r R_a^r(N_a) + R_a^r(N_a) A_a^{rT} + B_a^r B_a^{rT} = 0. \tag{2.96}$$

For the disturbance rejection problem with nonlinear sensor of Figure 2.11(b), $N_a = 1$ and $F_{\Omega_r}(s) = 0$. Thus, the matrices (2.85) become

$$A_s^d = \begin{bmatrix} A_p & B_p C_c & B_p C_d \\ -B_c N_s C_p & A_c & 0 \\ 0 & 0 & A_d \end{bmatrix},$$

$$B_s^d = \begin{bmatrix} 0 \\ 0 \\ B_d \end{bmatrix}, \tag{2.97}$$

$$C_s^d = \begin{bmatrix} 0 & C_c & 0 \\ C_p & 0 & 0 \end{bmatrix},$$

and the *disturbance rejection quasilinear gain equation* is

$$N_s = \mathcal{G}\left(\sqrt{C_{2s}^d R_s^d(N_s) C_{2s}^{dT}} \right), \tag{2.98}$$

where

$$C_{2s}^d = \begin{bmatrix} C_p & 0 & 0 \end{bmatrix} \tag{2.99}$$

and $R_s^d(N_s)$ is a positive definite solution of

$$A_s^d R_s^d(N_s) + R_s^d(N_s) A_s^{dT} + B_s^d B_s^{dT} = 0. \tag{2.100}$$

All other special cases of Subsections 2.2.3–2.2.5 can be obtained from (2.85)–(2.92) similarly.

2.3 Accuracy of Stochastic Linearization in Closed Loop LPNI Systems

In this section, we quantify the accuracy of stochastic linearization using both analytical and numerical techniques.

2.3.1 Fokker-Planck Equation Approach

Consider the disturbance rejection problem in the LPNI system shown in Figure 2.8(a). Assume that

$$P(s) = \frac{1}{s+1}, \ C(s) = 1, \ F_{\Omega_d}(s) = 1, \ f(\cdot) = \mathrm{sat}(\cdot). \tag{2.101}$$

Clearly, the behavior of this system is described by the following differential equation:

$$\dot{y} = -y + w - \mathrm{sat}(y). \qquad (2.102)$$

Our goal is to calculate the steady state standard deviation of y. Since the system is one-dimensional, this can be accomplished analytically using the Fokker-Planck equation. Indeed, for (2.102) the stationary Fokker-Planck equation is

$$\frac{1}{2}\frac{d^2\pi(y)}{dy^2} + [y + \mathrm{sat}(y)]\frac{d\pi(y)}{dy} + [1 + \mathrm{sat}'(y)]\pi(y) = 0, \qquad (2.103)$$

where $\pi(y)$ is the steady state pdf of y. The solution of (2.103) is given by

$$\pi(y) = \kappa \, \exp\left[-y^2 - 2\,\Phi(y)\right], \qquad (2.104)$$

where

$$\Phi(y) = \int_0^y f(u)\,du = \begin{cases} y^2/2, & |y| \le 1, \\ |y| - 1/2, & |y| > 1, \end{cases} \qquad (2.105)$$

and the normalization constant κ is calculated, using

$$\int_{-\infty}^{+\infty} \pi(y)\,dy = 1, \qquad (2.106)$$

to be $\kappa = 0.7952$. From the pdf (2.104), the standard deviation of y is

$$\sigma_y = 0.5070.$$

Now, we estimate this standard deviation using the quasilinear system of Figure 2.8(b). For this system with functional blocks defined by (2.101), the quasilinear gain equation (2.52) becomes

$$x - \mathrm{erf}\left(\frac{1}{\sqrt{2}\left\|\frac{1}{s+1+x}\right\|_2}\right) = 0.$$

Its solution, using Bisection Algorithm 2.1, is determined to be $x = 0.9518$. Thus, $N_a = 0.9518$ and

$$\sigma_{\hat{y}} = \left\|\frac{1}{s + 1.9518}\right\|_2 = 0.5061.$$

Clearly, the error between σ_y and $\sigma_{\hat{y}}$, defined as

$$\varepsilon = \frac{|\sigma_y - \sigma_{\hat{y}}|}{\sigma_y} \times 100\%, \qquad (2.107)$$

is 0.18%.

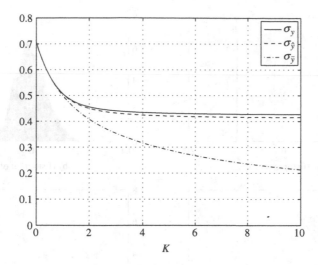

Figure 2.23. Standard deviations of y, $\hat{y}$, and $\tilde{y}$ as functions of the controller gain K.

To further characterize the accuracy of stochastic linearization, consider the system with all elements as in (2.101), but $C(s) = K \in (0, 10]$, and calculate σ_y and $\sigma_{\hat{y}}$ using the Fokker-Planck equation and stochastic linearization, respectively. The result is shown in Figure 2.23. Clearly, for all K, the accuracy is quite high, with a maximum error of 2.81%. For comparison, Figure 2.23 also shows the standard deviation $\sigma_{\tilde{y}}$ obtained using the Jacobian linearization. The maximum error in this case is about 50%.

2.3.2 Filter Hypothesis Approach

In this subsection, we show by simulations and a statistical experiment that low-pass filtering properties of the plant bring about *Gaussianization* of the signal at the plant output, which leads to high accuracy of stochastic linearization.

Simulation approach: Consider the reference tracking LPNI system shown in Figure 2.5, with all elements as in (2.101) but

$$f(u) = \text{sat}_{10}(u).$$

Using the reference tracking quasilinear gain equation (2.50), we determine that $N = 0.076$. This indicates that the actuator saturates practically always. This is confirmed by simulations. Indeed, simulating this system for a sufficiently long time, we obtain the histogram of the signal at the output of the saturation, v, as shown in Figure 2.24(a). The histogram of the signal at the output of the plant, y, is shown in Figure 2.24(b). Clearly, the filtering properties of the plant "Gaussianizes" the distribution of y. The pdf's of $\hat{v}$ and $\hat{y}$ are shown in Figures 2.24(c) and (d), respectively. Note that while the distributions of v and $\hat{v}$ are quite different, the distributions of y and $\hat{y}$ are quite similar.

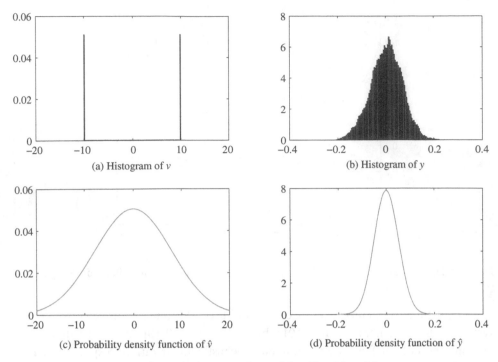

(a) Histogram of v

(b) Histogram of y

(c) Probability density function of $\hat{v}$

(d) Probability density function of $\hat{y}$

Figure 2.24. Histograms of v and y in comparison with pdf's of $\hat{v}$ and $\hat{y}$.

Statistical experiment approach: To further illustrate the accuracy of stochastic linearization, the following Monte Carlo experiment is performed: We consider 5000 disturbance rejection LPNI systems of Figure 2.8(a) with

$$C(s) = K, f(\cdot) = \text{sat}_\alpha(\cdot).$$

In 2000 of these systems we assume that

$$P(s) = \frac{1}{Ts+1} \tag{2.108}$$

and in the remaining 3000,

$$P(s) = \frac{\omega_n^2}{s^2 + 2\zeta\omega_n s + \omega_n^2}. \tag{2.109}$$

The parameters of these systems are randomly and equiprobably selected from the following sets:

$$T \in [0.01, 10], \ \omega_n \in [0.01, 10], \ \zeta \in [0.05, 1], \ K \in [1, 20], \ \alpha \in (0.1, 1]. \tag{2.110}$$

For each of these systems, we evaluate σ_y and $\sigma_{\hat{y}}$ using simulations and stochastic linearization, respectively. The accuracy is quantified by (2.107), the histogram of which is shown in Figure 2.25. Clearly, accuracy is high: 71.4% of the systems yield $\varepsilon < 5\%$ and only 9.2% of systems yield $\varepsilon > 10\%$. Further analysis reveals that these latter cases occur when the signals u and y are highly non-Gaussian, when either

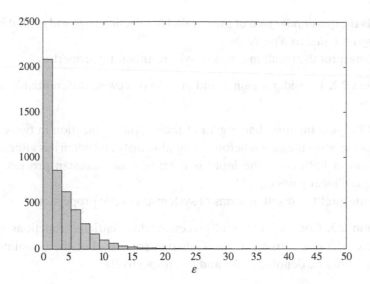

Figure 2.25. Histogram of ε for the Monte Carlo accuracy experiment.

$T \ll 1$ or $\zeta \ll 1$ (i.e., when low-pass filtering does not take place and, therefore, Gaussianization does not occur).

2.4 Summary

- Stochastic linearization of an isolated odd nonlinear function is the expected value of its derivative with respect to the Gaussian distribution of its argument. As a result, unlike the Jacobian linearization, stochastic linearization represents a nonlinear function globally.
- Stochastically linearized gains of actuators and sensors in closed loop LPNI systems are defined by transcendental equations, which can be solved using a bisection algorithm.
- The stochastically linearized gains of actuators and sensors depend not only on the shape of the nonlinearities involved but also on all other elements of the closed loop LPNI system, including the functional blocks and exogenous signals.
- Using its stochastically linearized version, standard deviations of all signals in a closed loop LPNI system can be evaluated.
- If the plant of an LPNI system is low-pass filtering, accuracy of stochastic linearization is quite high – well within 10% as far as the standard deviation of the output is concerned.

2.5 Problems

Problem 2.1. Consider two odd piecewise differentiable nonlinear functions $f_1(u)$ and $f_2(u)$ and assume that $u(t)$ is a zero-mean wss Gaussian process. Let the quasilinear gains of these functions be denoted as N_1 and N_2, respectively.

    (a) Is the quasilinear gain of the parallel connection of f_1 and f_2, that is, $f_1(u) + f_2(u)$, equal to $N_1 + N_2$?

    (b) Interpret the result in terms of system-theoretic properties.

Problem 2.2. Consider a gain k and an odd piecewise differentiable nonlinear function $f(\cdot)$.

    (a) Compute the quasilinear gain of their serial connection in two configurations: when the gain is before the nonlinearity and when it is after, assuming that in both cases the input is a zero-mean Gaussian process. Are the quasilinear gains equal?

    (b) Interpret the result in terms of system-theoretic properties.

Problem 2.3. Consider two odd piecewise differentiable functions $f_1(u)$ and $f_2(u)$, where $u(t)$ is a zero-mean wss Gaussian process. Let the quasilinear gains of these functions be denoted as N_1 and N_2, respectively.

    (a) Is the quasilinear gain of the serial connection of f_1 and f_2, that is, $f_2(f_1(u))$, equal to $N_1 N_2$? (Give "physical" arguments why this may or may not be true.)

    (b) If this is, in general, not true, find a class of functions for which this property holds.

    (c) Derive a general expression for the quasilinear gain of $f_2(f_1(u))$.

Problem 2.4. Consider the saturation function $\mathrm{sat}_\alpha^\beta(u(t))$, where α is the level of saturation, β is the slope of the linear part, and $u(t)$ is a zero-mean wss Gaussian process (see Figure 2.26).

    (a) For $\alpha = 1$ and $\beta = 5$, calculate and plot the quasilinear gain of this nonlinearity as a function of the standard deviation σ_u. Compare the result with that of Table 2.1.

    (b) For $\alpha = 1$ and $\sigma_u = 1$, calculate and plot the quasilinear gain as a function of β.

    (c) Formulate your conclusion as to the effect of β.

Problem 2.5. Consider the deadzone function $\mathrm{dz}_\Delta^\beta(u(t))$, where, as before, Δ is a half of the deadzone, β is the slope of the linear part, and $u(t)$ is a zero-mean wss Gaussian process (see Figure 2.27).

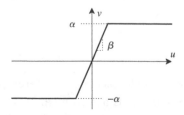

Figure 2.26. Problem 2.4.

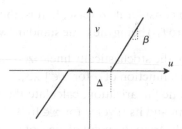

Figure 2.27. Problem 2.5.

   (a) For $\Delta = 1$ and $\beta = 5$, calculate and plot the quasilinear gain of this nonlinearity as a function of the standard deviation σ_u. Compare the result with that of Table 2.1.

   (b) For $\Delta = 1$ and $\sigma_u = 2$, calculate and plot the quasilinear gain as a function of β.

   (c) Formulate your conclusion as to the effect of β.

Problem 2.6. Consider an odd piecewise differentiable nonlinearity $f(u)$, where $u(t)$ is a zero-mean wss Gaussian process, and let $N(\sigma)$ be its quasilinear gain.

   (a) Show that

$$N(0) = f'(0)$$

   (b) Do you expect a similar relation hold away from the origin?

Problem 2.7. Consider an odd nonlinearity $f(u)$, where $u(t)$ is a zero-mean wss Gaussian process. Assume that $f(u)$ is linear in the vicinity of the origin, that is, $f(u) = mu$ for $|u| \leq \Delta$, and let $N(\sigma)$ be its quasilinear gain. Show that

$$N(\sigma) \approx m$$

in the vicinity of the origin, that is, on an interval whose length is of the order Δ.

Problem 2.8. Consider a sector bounded nonlinearity $f(u)$, where $u(t)$ is a zero-mean wss Gaussian process, and let $N(\sigma)$ be its quasilinear gain. Show that the latter is bounded by the magnitude of the sector, that is, for all $\sigma \geq 0$,

$$N(\sigma) < \sup_{u>0} \left| \frac{f(u)}{u} \right|.$$

Problem 2.9. Consider the serial connection of two blocks, where the first one is a linear dynamical system with transfer function

$$G(s) = \frac{1}{s+1}$$

and the second is defined by

$$f(\cdot) = \mathrm{sat}_\alpha(\cdot).$$

The input to this system is a standard white Gaussian process.

Along with this system consider its *reverse*, that is, the first block is $f(\cdot)$ and the second $G(s)$, with the input to $f(\cdot)$ being the same standard white Gaussian process.

(a) Calculate and plot the stochastically linearized gains of the original system and its reverse as a function of α for $\alpha \in [0,3]$.

(b) Using the stochastic linearization, calculate the variance at the output of the original system and its reverse for $\alpha \in [0,3]$.

(c) Interpret the results from the point of view of the effects of the nonlinearity and its position in this system (i.e., either before or after $G(s)$).

Problem 2.10. Figure 2.3 shows how the quasilinear gain of the saturation nonlinearity in Figure 2.2(a) depends on its authority α when the coloring filter and controller are given in (2.45).

(a) For the coloring filter

$$F_{\Omega_r} = \frac{4}{s+8}$$

and the controllers listed below, draw a plot, similar to Figure 2.3, showing how the quasilinear gain of the saturation nonlinearity depends on α.

(i) $C(s) = 1$;

(ii) $C(s) = \dfrac{1}{s+1}$;

(iii) $C(s) = \dfrac{1}{0.1s+1}$.

(b) Interpret how the time constant of the controller affects the quasilinear gain of the nonlinearity.

Problem 2.11. Consider the system of Figure 2.5(a) with coloring filter $F_{\Omega_r}(s)$ given in (2.45) and with plant, controller, and nonlinearity given below:

(i) $P(s) = \dfrac{1}{s(s+1)}, C(s) = 5, f(u) = \mathrm{sat}_\alpha(u)$;

(ii) $P(s) = \dfrac{1}{s(0.1s+1)}, C(s) = 1, f(u) = \mathrm{sat}_\alpha(u)$;

(iii) $P(s) = \dfrac{1}{s(0.01s+1)}, C(s) = 5, f(u) = \mathrm{sat}_\alpha(u)$.

(a) For each instance, draw a plot showing how the quasilinear gain behaves as a function of the authority of the saturation.

(b) Interpret how the time constant of the plant and the gain of the controller affect the quasilinear gain of the nonlinearity.

Problem 2.12. Consider the system of Figure 2.28, where $r(t)$ is a zero-mean wss Gaussian process and

$$P_1(s) = \frac{1}{s+1}, \; P_2(s) = \frac{2}{s+1}, \; C(s) = \frac{3}{s+3}, \; f(u) = \mathrm{sat}_\alpha(u).$$

For this system, compute the quasilinear gain of the nonlinearity and plot it as a function of α.

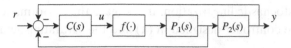

Figure 2.28. Problem 2.12.

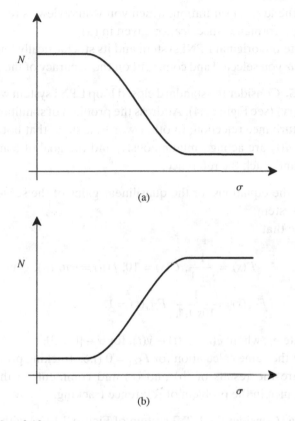

Figure 2.29. Problem 2.13.

Problem 2.13. Sketch the nonlinearities whose quasilinear gains are shown in Figures 2.29(a) and 2.29(b).

Problem 2.14. Consider the disturbance rejection closed loop LPNI system of Figure 2.8(a) with the plant

$$P(s) = \frac{1}{s+1},$$

the actuator

$$f(u) = \text{sat}_\alpha(u)$$

and the disturbance being standard white Gaussian process.

(a) Ignoring the actuator saturation, design a static linear controller that results in the standard deviation of the output of this system $\sigma_y = 0.1$.

(b) Using stochastic linearization, investigate the performance of this system with the controller designed in (a) and the saturating actuator having $\alpha \in [0.1, 3]$.

(c) Evaluate the performance degradation due to saturation for the α's investigated.

(d) Select the level of saturation, which you would view as reasonable in light of the performance specification given in (a).

(e) Simulate the original LPNI system and its stochastically linearized version for the α you selected and comment on the accuracy of the approximation.

Problem 2.15. Consider the standard closed loop LPNI system with the actuator $f(u)$ and sensor $g(y)$ (see Figure 2.4). Address the problem of simultaneous reference tracking and disturbance rejection. In other words, assume that both reference $r(t)$ and disturbance $d(t)$ are acting simultaneously, and the goal of control is to reject the disturbance and track the reference.

(a) Derive the equations for the quasilinear gains of the sensor and actuator in this system.

(b) Assume that

$$P(s) = \frac{1}{s+1}, \ C(s) = 10, \ f(u) = \text{sat}_\alpha(u), \ g(y) = y,$$

$$F_{\Omega_r}(s) = \frac{1}{10s+1}, \ F_{\Omega_d}(s) = 1.$$

Evaluate $\sigma_{\hat{e}}$, where $\hat{e}(t) = r(t) - \hat{y}(t)$, for $\alpha \in [0.1, 3]$.

(c) Repeat the same calculation for $F_{\Omega_d} = 0$ (i.e., tracking problem only).

(d) Compare the results of (b) and (c) and comment on the effect of the disturbance in the problem of reference tracking.

Problem 2.16. Consider the LPNI system of Figure 2.14(a) with

$$P(s) = \frac{1}{s+1}, \ C(s) = 1, \ F_{\Omega_r}(s) = \frac{\sqrt{3}}{s^3 + 2s^2 + 2s + 1},$$

$$f(u) = \text{sat}_\alpha(u), \ g(y) = \text{dz}_\Delta(y).$$

(a) Using the elimination procedure of (2.70), compute the quasilinear gains of the two nonlinearities.

(b) For a fixed value of α, plot the quasilinear gains as functions of Δ.

(c) For a fixed value of Δ, plot the quasilinear gains as functions of α.

Problem 2.17. Demonstrate the Gaussianization property of filtering plants. To accomplish this:

(a) Select any sufficiently filtering transfer function.

(b) Select the input of this transfer function as any non-Gaussian stationary random process; plot the histogram of this process.

(c) Simulate this (open loop) system and plot the histogram of the output process; observe a "more Gaussian" nature of this histogram as compared with the one at the input.

(d) Repeat the same procedure for a less filtering transfer function and observe that the Gaussianization takes place to a lesser extent (if any).

Problem 2.18. This problem is intended to investigate stochastic linearization of non-odd nonlinearities. Specifically, assume that the actuator is given by (see Figure 2.30)

$$\text{sat}_{\alpha\beta}(u) = \begin{cases} \alpha, & u > \alpha, \\ u, & -\beta \leq u \leq \alpha, \\ -\beta, & u < -\beta. \end{cases}$$

(a) Derive the expression for the quasilinear gain of $\text{sat}_{\alpha\beta}(u)$, where $u(t)$ is zero-mean wss Gaussian process.

(b) Plot the quasilinear gain as a function of α for $\beta = 1$.

(c) Plot the quasilinear gain as a function of β for $\alpha = 1$.

(d) Interpret the results.

Problem 2.19. Investigate the performance of LPNI system with non-odd actuator in the problem of reference tracking. For this reason, consider the system of Figure 2.5(a) with

$$P(s) = \frac{10}{s+10}, \quad C(s) = 5, \quad F_{\Omega_r}(s) = \frac{\sqrt{3}}{s^3 + 2s^2 + 2s + 1}$$

and $f(\cdot)$ given in Figure 2.30 with $\alpha = 2$ and $\beta = 1$.

(a) Evaluate the expected value and standard deviation of the error signal in the stochastically linearized system (i.e., $E(\hat{e}), \sigma_{\hat{e}}$).

(b) Compare the result with that obtained in Example 2.2.

Problem 2.20. Investigate the performance of LPNI system with non-odd actuator in the problem of disturbance rejection. For this reason, consider the system of Figure 2.8(a) with

$$P(s) = \frac{10}{s+10}, \quad C(s) = 5, \quad F_{\Omega_d}(s) = \frac{\sqrt{3}}{s^3 + 2s^2 + 2s + 1}$$

and $f(\cdot)$ given in Figure 2.30 with $\alpha = 2$ and $\beta = 1$.

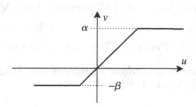

Figure 2.30. Problem 2.18.

(a) Evaluate the expected value and standard deviation of the error signal in the stochastically linearized system (i.e., $E(\hat{y}), \sigma_{\hat{y}}$).

(b) Compare the result with that obtained in Example 2.3.

2.6 Annotated Bibliography

The idea of stochastic linearization appeared, practically simultaneously, in the following two articles:

[2.1] R.C. Booton, "Nonlinear control systems with random inputs," *IRE Transactions on Circuit Theory*, Vol. CT-1, pp. 9–18, 1954.

[2.2] I.E. Kazakov, "Approximate method for the statistical analysis of nonlinear systems," *Technical Report VVIA 394*, Trudy, 1954 (in Russian)

The main monographs devoted to the method of stochastic linearization in the framework of feedback systems are discussed in the following:

[2.3] I.E. Kazakov and B.G. Dostupov, *Statistical Dynamics of Nonlinear Automatic Systems*, Fizmatgiz, Moscow, 1962 (in Russian)

[2.4] A.A. Pervozvansly, *Random Processes in Nonlinear Automatic Systems*, Fizmatgiz, Moscow, 1962 (in Russian)

[2.5] A. Gelb and W.E. Van der Velde, *Multiple Input Describing Functions and Nonlinear Design*, McGrow-Hill, New York, 1968

In the framework of mechanical systems, this method and its numerous applications are described in the following:

[2.6] J.B. Roberts and P.D. Spanos, *Random Vibrations and Statistical Linearization*, John Wiley and Sons, Chichester, 1990 (Dover edition 1999)

For problems in Physics, this method is further developed in the following:

[2.7] L. Socha, *Linearization Methods for Stochastic Dynamic Systems*, Springer, Berlin, 2008

As one can see from the titles of these monographs, the method has been referred to as either *statistical linearization* or *random describing functions*. We prefer to call it *stochastic linearization* since it involves no "statistics" in the sense of the current usage of this term and since describing function arc strongly associated with periodic regimes in feedback systems. Note that the term *stochastic linearization* has also been used in the past, for example in the following:

[2.8] W.M. Wonham and W.F. Cashman, "A computational approach to optimal control of stochastic saturating systems," *International Journal of Control*, Vol. 10, pp. 77–498, 1969

[2.9] I. Elishakoff, "Stochastic linearization technique: A new interpretation and a selective review," *The Shock and Vibration Digest*, Vol. 32, pp. 179—-188, 2000

Relationship (2.8) used in the Proof of Theorem 2.1 can be found in the following:

[2.10] A. Papoulis, *Probability, Random Variables and Stochastic Processes*, McGrow-Hills, New York, 1984

Main results of large deviations theory can be found in the following:

[2.11] M.I. Freidlin and A.D. Ventzell, *Random Perturbations of Dynamical Systems*, Springer, New York, 1984

[2.12] Z. Shuss, *Theory and Applications of Stochastic Differential Equations*, Springer, New York, 2009.

There have been numerous attempts to characterize the accuracy of stochastic linearization. However, no general results in this direction (similar to those for describing functions and harmonic balance) have been obtained. Some special cases have been investigated and high accuracy has been demonstrated in [2.5]–[2.9] and also in the following:

[2.13] J. Skrzypczyk, "Accuracy analysis of statistical linearization methods applied to nonlinear dynamical systems," *Reports on Mathematical Physics*, Vol. 36, pp. 1–20, 1995

[2.14] C. Gökçek, P.T. Kabamba, and S.M. Meerkov, "Disturbance rejection in control systems with saturating actuators," *Nonlinear Analysis*, Vol. 40, pp. 213–226, 2000

[2.15] S. Ching, S.M. Meerkov, and T. Runolfsson, "Gaussianization of random inputs by filtering plants: The case of poisson white and telegraph processes," Proceedings of the 2010 Conference on Decision and Control, Atlanta, GA, 2010

3 Analysis of Reference Tracking in LPNI Systems

Motivation: This chapter is intended to quantify steady state errors and transient performance of LPNI systems in the problem of tracking reference signals.

For linear systems, the steady state error in tracking deterministic references is characterized by the notion of system type, defined by the poles of the loop gain at the origin. This definition is not applicable to LPNI systems since, if the actuator saturates, controller poles at the origin play a different role than those of the plant and, thus, the loop gain does not define the system type. This motivates the first goal of this chapter, which is to introduce the notion of system type for LPNI systems.

Also in contrast with the linear case, LPNI systems with saturating actuators are not capable of tracking steps with arbitrary amplitudes. This motivates the second goal of this chapter: to introduce and quantify the notion of trackable domains, that is, the sets of step (or ramp, or parabolic) input magnitudes that can, in fact, be tracked by an LPNI system.

As far as the transients are concerned, the usual step tracking measures, such as overshoot, settling time, and so on, are clearly not appropriate for random references. The variance of tracking errors, as it turns out, is not a good measure either, since for the same variance, the nature of tracking errors can be qualitatively different. This motivates the third goal of this chapter: to introduce novel transient tracking quality indicators and utilize them for analysis of random reference tracking in both linear and LPNI closed loop systems.

Overview: For deterministic references, the notions of LPNI system type and trackable domain are introduced and analyzed in Section 3.1. For random references, the notions of random sensitivity function and saturating random sensitivity function as well as tracking quality indicators are introduced in Sections 3.2 and 3.3, respectively. It is shown that the tracking quality indicators can be used for LPNI systems analysis and design in the same manner as settling time and overshoot are used in the linear case.

3.1 Trackable Domains and System Types for LPNI Systems

3.1.1 Scenario

The material of this section is somewhat different from the rest of this book in that it does not deal with random reference signals and does not use the method of stochastic linearization. Nevertheless, it is included here for two reasons: First, the notions of LPNI system type and trackable domain, introduced and investigated here, are useful by themselves when one deals with systems having saturating and other nonlinear actuators. Second, the notion of trackable domain is used elsewhere in this volume (including later on in this chapter) to characterize tracking qualities of closed loop LPNI systems with random reference signals.

It is intuitively clear that the notion of system type in LPNI systems must depend on the nonlinearities involved. Indeed, saturating actuators and actuators with dead-zone respond to step inputs in a qualitatively different manner, leading to different notions of system type. Therefore, to be specific, we center here on LPNI systems with a ubiquitous nonlinearity – saturating actuators.

In LPNI systems with saturating actuators, loop gain does not define the quality of reference tracking. Indeed, consider, for example, the two systems of Figure 3.1, which have identical loop gain and system type if the saturation is ignored. However, in the presence of saturation, they have qualitatively different tracking capabilities. Namely, the system of Figure 3.1(a) can track ramps with a finite steady state error, but that of Figure 3.1(b) cannot track ramps at all (see Figure 3.2). Thus, the notion of system type has to be modified, and this is carried out in this section.

Specifically, we show that the LPNI system type is defined by only plant poles. The controller poles, however, also play a role but it is limited to affecting the steady state errors, while not enlarging the class of trackable references. This class is characterized by a new notion of *trackable domain*, which quantifies the size of steps or the slope of ramps, that can be tracked when saturation is present.

3.1.2 Trackable Domains and Steady State Errors

Step input: Consider the reference tracking LPNI system of Figure 3.3, where $P(s)$ and $C(s)$ are the plant and the controller, respectively, and $\text{sat}_\alpha(\cdot)$ is the saturating

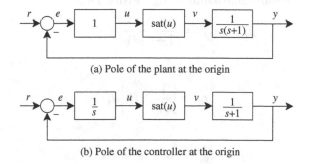

(a) Pole of the plant at the origin

(b) Pole of the controller at the origin

Figure 3.1. Motivating example.

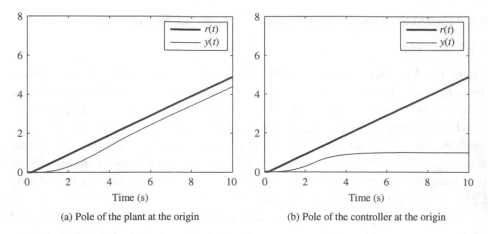

(a) Pole of the plant at the origin   (b) Pole of the controller at the origin

Figure 3.2. Ramp tracking by systems of Figure 3.1.

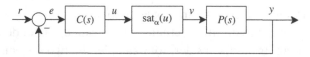

Figure 3.3. LPNI system with saturating actuator.

actuator of authority α. Assume that

$$r(t) = r_0 \mathbf{1}(t), \tag{3.1}$$

where $\mathbf{1}(t)$ is the unit step function and $r_0 \in \mathbf{R}$. Define the steady state tracking error:

$$e_{ss}^{step} = \lim_{t \to \infty} [r_0 \mathbf{1}(t) - y(t)]. \tag{3.2}$$

Let C_0 and P_0 be the d.c. gains of the controller and plant, respectively, that is,

$$C_0 = \lim_{s \to 0} C(s), \tag{3.3}$$

$$P_0 = \lim_{s \to 0} P(s). \tag{3.4}$$

Theorem 3.1. *Assume that a unique e_{ss}^{step} exists. Then,*

(i) $$e_{ss}^{step} = \frac{r_0}{1 + C_0 P_0}, \tag{3.5}$$

if

$$|r_0| < \left| \frac{1}{C_0} + P_0 \right| \alpha; \tag{3.6}$$

(ii) $$e_{ss}^{step} = r_0 - (\text{sign } r_0 C_0) P_0 \alpha, \tag{3.7}$$

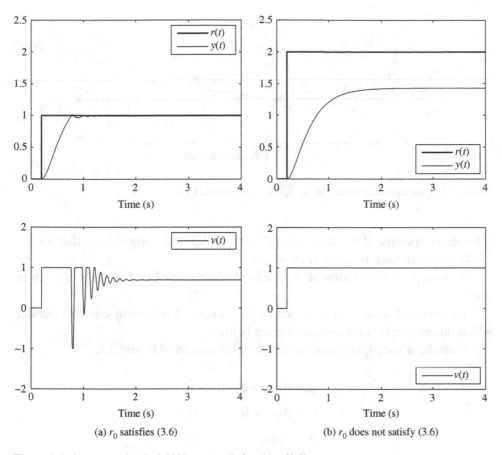

Figure 3.4. Step tracking in LPNI system defined by (3.9).

if

$$|r_0| \geq \left| \frac{1}{C_0} + P_0 \right| \alpha. \tag{3.8}$$

Proof. See Section 8.2

Thus, if r_0 satisfies (3.6), e_{ss}^{step} is the same as in systems with linear actuators. However, if r_0 is outside of (3.6), that is, in (3.8), no tracking takes place since $e_{ss}^{step} = r_0 -$ const. This is illustrated in Figure 3.4 (along with the signal $v(t)$ at the output of the actuator) for the system of Figure 3.3 with

$$P(s) = \frac{30}{s^2 + 10s + 21}, \quad C(s) = 100, \quad \alpha = 1. \tag{3.9}$$

The set of r_0 defined by (3.6) is referred to as the *Step Trackable Domain* or just *Trackable Domain (TD)*. Clearly, for $P_0 > 0$, $C_0 > 0$, the size of *TD*, that is, $|TD|$, is a monotonically increasing function of P_0 and α and a monotonically decreasing function of C_0. If $P(s)$ has at least one pole at the origin, $|TD| = \infty$; otherwise, $|TD|$

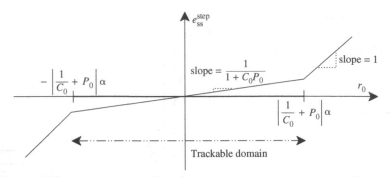

Figure 3.5. Conceptual illustration of e_{ss}^{step} as a function of r_0.

is finite, irrespective of whether or not $C(s)$ has poles at the origin. Note that for the LPNI system defined by (3.9), $|TD| = 1.44$.

A conceptual illustration of the trackable domain in LPNI systems is given in Figure 3.5.

Theorem 3.1 assumes that a unique e_{ss}^{step} exists. A sufficient condition under which this assumption takes place is given below:

Consider a state space representation of the system of Figure 3.3:

$$
\begin{aligned}
\dot{x}_p &= A_p x_p + B_p \text{sat}_\alpha(u), \\
\dot{x}_c &= A_c x_c + B_c(r_0 - y), \\
y &= C_p x_p, \\
u &= C_c x_c + D_c(r_0 - y),
\end{aligned}
\tag{3.10}
$$

where x_p and x_c are the states of the plant and the controller, respectively.

Theorem 3.2. *Assume that with $r_0 = 0$, the equilibrium point $x_p = 0$, $x_c = 0$ of (3.10) is globally asymptotically stable, and this fact can be established by means of a Lyapunov function of the form*

$$
V(x) = x^T Q x + \int_0^{Cx} \text{sat}_\alpha(\tau) d\tau, \quad Q = Q^T \geq 0,
\tag{3.11}
$$

where $C = [-D_c C_p \ C_c]$ and $x = [x_p^T \ x_c^T]^T$. Then, for $r_0 \in TD$, e_{ss}^{step} exists and is unique.

Proof. See Section 8.2

As an example, consider the system of Figure 3.3 with

$$
P(s) = \frac{1}{s(s+1)}, \quad C(s) = 10, \quad \alpha = 1,
\tag{3.12}
$$

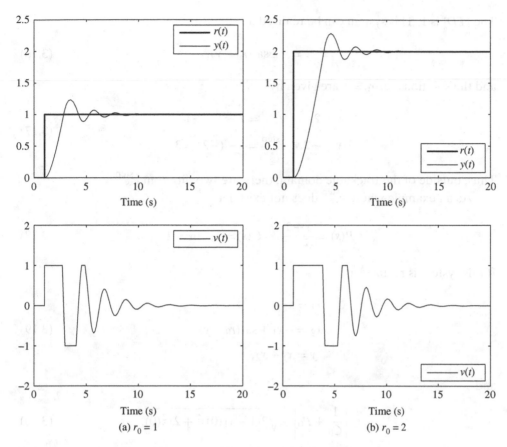

Figure 3.6. Step tracking in LPNI system defined by (3.12).

and the following realization:

$$A_p = \begin{bmatrix} 0 & 1 \\ 0 & -1 \end{bmatrix}, \quad B_p = \begin{bmatrix} 0 \\ 1 \end{bmatrix}, \quad C_p = \begin{bmatrix} 1 & 0 \end{bmatrix},$$

$$A_c = 0, \ B_c = 0, \ C_c = 0, \ D_c = 10.$$

(3.13)

Then the condition in Theorem 3.2 is satisfied with

$$Q = \begin{bmatrix} 0 & 0 \\ 0 & 5 \end{bmatrix}.$$

(3.14)

Thus, according to Theorem 3.2 a unique e_{ss}^{step} does exist. In addition, Theorem 3.1 ensures that $e_{ss}^{step} = 0$ for any r_0. This is illustrated in Figure 3.6.

Outside the trackable domain, e_{ss}^{step} may either be not unique or not exist at all. Examples for each of these cases are as follows:

For the case of nonuniqueness, consider the system of Figure 3.3 with

$$P(s) = -2, \ C(s) = 1, \ \alpha = 1.$$

(3.15)

The $|TD|$ is 1. This system can be realized as

$$y = -2\,\text{sat}(r_0 - y),\tag{3.16}$$

and the solutions for $r_0 = 1$ are given by

$$\begin{aligned}y &= 2, & e_{\text{ss}}^{\text{step}} &= 1 - 2 = -1\\ y &= -2, & e_{\text{ss}}^{\text{step}} &= 1 - (-2) = 3\end{aligned}\tag{3.17}$$

Thus, outside of the trackable domain, there are two different $e_{\text{ss}}^{\text{step}}$.
As an example where $e_{\text{ss}}^{\text{step}}$ does not exist, let

$$P(s) = \frac{s+1}{s^2+1}, \quad C(s) = 1, \quad \alpha = 1.\tag{3.18}$$

If this system is realized as

$$\begin{aligned}\dot{x}_1 &= x_2,\\ \dot{x}_2 &= -x_1 + \text{sat}(r_0 - y),\\ y &= x_1 + x_2,\end{aligned}\tag{3.19}$$

and

$$r_0 \geq \left|\frac{1}{C_0} + P_0\right| + \sqrt{2(1 - x_1(0))^2 + 2x_2^2(0)},\tag{3.20}$$

the solution is given by

$$\begin{aligned}x_1(t) &= 1 - (1 - x_1(0))\cos t + x_2(0)\sin t,\\ x_2(t) &= (1 - x_1(0))\sin t + x_2(0)\cos t,\end{aligned}\tag{3.21}$$

implying that $e_{\text{ss}}^{\text{step}}$ does not exist.

Ramp and parabolic inputs: Consider the system of Figure 3.3 with

$$r(t) = r_1 t\mathbf{1}(t)\tag{3.22}$$

or

$$r(t) = \frac{1}{2}r_2 t^2 \mathbf{1}(t)\tag{3.23}$$

and define steady state errors as

$$e_{\text{ss}}^{\text{ramp}} = \lim_{t\to\infty}[\,r_1 t\mathbf{1}(t) - y(t)\,],\tag{3.24}$$

$$e_{\text{ss}}^{\text{par}} = \lim_{t\to\infty}\left[\frac{r_2}{2}t^2\mathbf{1}(t) - y(t)\right].\tag{3.25}$$

Let C_0, as before, be the d.c. gain of the controller and P_1 and P_2 be defined by

$$P_1 = \lim_{s \to 0} sP(s), \tag{3.26}$$

$$P_2 = \lim_{s \to 0} s^2 P(s). \tag{3.27}$$

Theorem 3.3. *The steady state errors for the ramp and parabolic inputs can be characterized as follows:*
(i) Assume a unique e_{ss}^{ramp} exists. Then,

$$e_{ss}^{ramp} = \frac{r_1}{C_0 P_1} \tag{3.28}$$

if

$$|r_1| < |P_1|\alpha. \tag{3.29}$$

(ii) Assume that a unique e_{ss}^{par} exists. Then,

$$e_{ss}^{par} = \frac{r_2}{C_0 P_2} \tag{3.30}$$

if

$$|r_2| < |P_2|\alpha. \tag{3.31}$$

Proof. See Section 8.2

The ranges of r_1 and r_2, defined by (3.29) and (3.31), are referred to as *Ramp* and *Parabolic Trackable Domains*, respectively. From (3.28)–(3.31), the following conclusions can be made:

- Unlike the case of step inputs, the expressions for the steady state errors (3.28) and (3.30) differ from those in systems with linear actuators.
- The ramp trackable domain is non-empty if and only if the plant has at least one pole at the origin. For the parabolic trackable domain to be non-empty, the plant must have at least two poles at the origin.
- Although the controller poles at the origin do not contribute to the size of the ramp and parabolic trackable domains, they play a role in the size of the steady state errors: if $C(s)$ has one or more poles at the origin, the steady state errors in the trackable domains are zero even if P_1 or P_2 are finite.

These conclusions are illustrated in Figure 3.7 for the LPNI system of Figure 3.3 with

$$P(s) = \frac{15}{s(s+10)}, \quad C(s) = 5, \quad \alpha = 1. \tag{3.32}$$

If $C(s)$ has a pole at the origin, the system of Figure 3.3 is sometimes augmented by an anti-windup compensator. The purpose of anti-windup is to improve transient response and stability while maintaining small signal ($|u(t)| < \alpha$) behavior identical

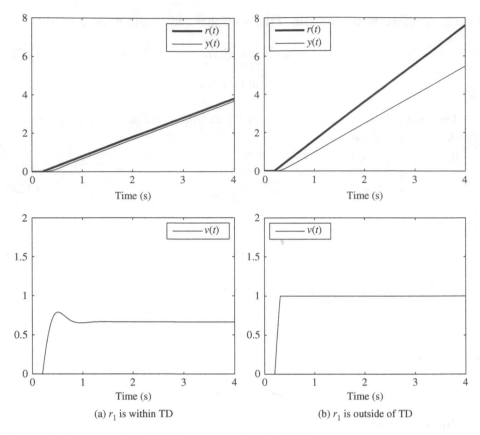

Figure 3.7. Ramp tracking in LPNI system defined by (3.32).

to that of the closed loop system without anti-windup. Thus, if $u(t)$ in the system with anti-windup does not saturate at steady state, its asymptotic tracking properties remain the same as those of the system without anti-windup. The condition on r_0 in Theorem 3.1 (r_1, r_2 in Theorem 3.3) is sufficient to ensure $|\lim_{t\to\infty} u(t)| < \alpha$. Therefore, Theorems 3.1 and 3.3 remain valid in the presence of anti-windup as well. This fact holds regardless of the structure of anti-windup as long as the closed loop system with anti-windup maintains small signal ($|u(t)| < \alpha$) behavior identical to that of the closed loop system without anti-windup.

3.1.3 System Types

Theorems 3.1 and 3.3 show that the roles of $C(s)$ and $P(s)$ in systems with saturating actuators are different. Therefore, system type cannot be defined in terms of the loop transfer function $C(s)P(s)$, as it is in the linear case.

To motivate a classification appropriate for the saturating case, consider again the system of Figure 3.1(b). According to Theorem 3.1, this system tracks steps from $TD = \{r_0 \in \mathbf{R} : |r_0| < \alpha\}$, with zero steady state error. However, this system does not track any ramp input at all. Thus, this system is similar to a type 1 linear system but

Table 3.1. *Steady state errors and Trackable Domains for various system types*

| Type | $r(t) = r_0\mathbf{1}(t)$ | | $r(t) = r_1 t\mathbf{1}(t)$ | | $r(t) = \frac{1}{2}r_2 t^2\mathbf{1}(t)$ | |
	TD	e_{ss}^{step}	TD	e_{ss}^{ramp}	TD	e_{ss}^{par}
0_S	$\|r_0\| < \left\|\frac{1}{C_0} + P_0\right\|\alpha$	$\frac{r_0}{1+P_0C_0}$	$\varnothing$	Not applicable	$\varnothing$	Not applicable
0_S^+	$\|r_0\| < \|P_0\|\alpha$	0	$\varnothing$	Not applicable	$\varnothing$	Not applicable
1_S	$\mathbf{R}$	0	$\|r_1\| < \|P_1\|\alpha$	$\frac{r_1}{P_1C_0}$	$\varnothing$	Not applicable
1_S^+	$\mathbf{R}$	0	$\|r_1\| < \|P_1\|\alpha$	0	$\varnothing$	Not applicable
2_S	$\mathbf{R}$	0	$\mathbf{R}$	0	$\|r_2\| < \|P_2\|\alpha$	$\frac{r_2}{P_2C_0}$

not exactly: it does not track ramps. Therefore, an "intermediate" system type is necessary. This observation motivates the following definition.

Definition 3.1. *The system of Figure 3.3 is of type k_S, where S stands for saturating, if the plant has k poles at the origin. It is of type k_S^+ if, in addition, the controller has one or more poles at the origin.*

In terms of these system types, the steady state errors and trackable domains in systems with saturating actuators are characterized in Table 3.1. As it follows from Definition 3.1, the system of Figure 3.1(a) is of type 1_S, whereas that of Figure 3.1(b) is of type 0_S^+. This, together with Table 3.1, explains the difference in their tracking capabilities alluded to in Section 3.1.

Note that the definition of system type k_S is a proper extension of types for linear systems. Indeed, when the level of saturation, α, tends to infinity, all trackable domains become the real line $\mathbf{R}$ and type k_S systems become the usual type k.

In Table 3.1, only types up to 2_S are included since systems of higher types can only be stabilized locally by linear controllers with actuator saturation. Trackable domains and steady state errors for systems of type higher than 2_S can also be characterized if the assumption on global stability of the closed loop system (3.10) is replaced by local stability and initial conditions are restricted to the domain of attraction. This, however, would necessitate estimating domains of attraction, which is generally difficult. Moreover, the trackable domain will depend on this estimate. Therefore, we limit our analysis to globally asymptotically stable cases.

3.1.4 Application: Servomechanism Design

In this subsection, we illustrate the controller design process, using the saturating system types. Specifically, for a given plant with saturating actuator, we select a controller structure to satisfy steady state performance specifications and then adjust the controller parameters to improve transient response.

Consider a DC motor modeled by

$$P(s) = \frac{1}{s(Js+b)}, \tag{3.33}$$

where $J = 0.06$ kg $\cdot$ m^2 and $b = 0.01$ kg $\cdot$ m^2/s. Assume that the maximum attainable torque of the motor is 2.5 N $\cdot$ m, that is,

$$\alpha = 2.5. \tag{3.34}$$

Using this motor, the problem is to design a servomechanism satisfying the following steady state performance specifications:

(i) steps of magnitude $|r_0| < 100$ rad should be tracked with zero steady state error;

(ii) ramps of slope $|r_1| < 20$ rad/s should be tracked with steady state error less than or equal to $0.01r_1$.

A solution to this problem is as follows: According to Table 3.1, specification (i) requires a system of type at least 0_S^+. Since the plant has a pole at the origin, the system is of type at least 1_S, which guarantees $e_{ss}^{step} = 0$ and $TD = \mathbf{R}$. Specification (ii) requires

$$|P_1|\alpha \geq 20, \tag{3.35}$$

and

$$e_{ss}^{ramp} = \frac{r_1}{P_1 C_0} \leq 0.01r_1. \tag{3.36}$$

Inequality (3.35) is satisfied, since $P_1 = 100$ and $\alpha = 2.5$. Inequality (3.36) is met if $C_0 \geq 1$. Thus, a controller of the form

$$C(s) = C_0 \prod_{i=1}^{k} \frac{\tau_i s + 1}{T_i s + 1}, \quad k \geq 1, \ \tau_i \geq 0, \ T_i \geq 0 \tag{3.37}$$

guarantees (i) and (ii) if $C_0 \geq 1$.

Note that, at steady state, the output of the plant cannot grow faster than $|P_1|\alpha$ rad/s. Hence, if a ramp trackable domain were specified by $|r_1| < M$ rad/s with $M > |P_1|\alpha = 250$ rad/s, no linear controller, satisfying this specification, would exist.

Although the closed loop system (3.33), (3.34), and (3.37) does satisfy the steady state specifications, it is of interest to analyze its transient behavior as well. To accomplish this, consider the simplest controller of the form (3.37):

$$C(s) = 1. \tag{3.38}$$

The transients of (3.33), (3.34), and (3.38), shown in Figure 3.8, are deficient: both the overshoot and the settling time are too large. A similar situation takes place for ramp inputs as well: transients are too long. To improve this behavior, consider the controller (3.37) with $C_0 = 1$, $\tau_1 = 6.667$, and $T_1 = 0.02$, that is,

$$C(s) = \frac{6.667s + 1}{0.02s + 1}. \tag{3.39}$$

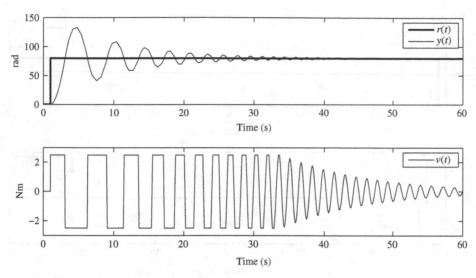

Figure 3.8. Transient response of (3.33) and (3.34) with $C(s) = 1$.

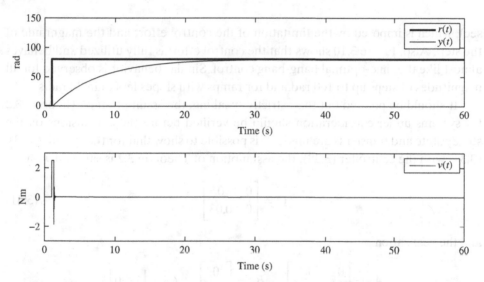

Figure 3.9. Transient response of (3.33) and (3.34) with $C(s) = \frac{6.667s+1}{0.02s+1}$.

The transients of (3.33), (3.34), and (3.39), illustrated in Figure 3.9, show an improvement in overshoot but still long settling time. This, perhaps, is due to the fact that, as it follows from Figure 3.9, the control effort is underutilized. To correct this situation, consider the controller (3.37) with $C_0 = 1$, $\tau_1 = 0.667$, and $T_1 = 0.02$, that is,

$$C(s) = \frac{0.667s+1}{0.02s+1}. \tag{3.40}$$

With this controller, the transients are shown in Figure 3.10. As one can see, the settling time has improved considerably and is very close to its lower limit (about 2.1

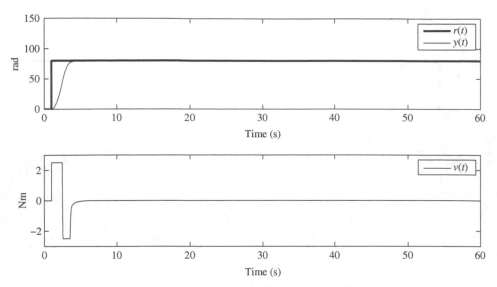

Figure 3.10. Transient response of (3.33) and (3.34) with $C(s) = \frac{0.667s+1}{0.02s+1}$.

sec), which is imposed by the limitation of the control effort and the magnitude of the step. Also, Figure 3.10 shows that the control effort is fully utilized and behaves almost like the time-optimal bang-bang control. Similar behavior is observed for all magnitudes of steps up to 100 rad and for ramps with slopes less than 20 rad/s.

It should be pointed out that, strictly speaking, the assumption of Theorem 3.2 for systems under consideration should be verified before the conclusions on the steady state and transients are made. It is possible to show that for the system (3.33), (3.34) with the controller (3.38), the assumption of Theorem 3.2 is satisfied with

$$Q = \begin{bmatrix} 0 & 0 \\ 0 & 0.03 \end{bmatrix} \tag{3.41}$$

and the realization

$$A_p = \begin{bmatrix} 0 & 1 \\ 0 & -b/J \end{bmatrix}, \quad B_p = \begin{bmatrix} 0 \\ 1/J \end{bmatrix}, \quad C_0 = \begin{bmatrix} 1 & 0 \end{bmatrix}, \tag{3.42}$$

$$A_c = 0, \ B_c = 0, \ C_c = 0, \ D_c = 1.$$

If the controller (3.39) is used,

$$Q = \begin{bmatrix} 0.25 & 1.5 & 0 \\ 1.5 & 10 & 0 \\ 0 & 0 & 0 \end{bmatrix} \tag{3.43}$$

and the realization

$$A_c = -50, \ B_c = -50(50/0.15 - 1), \ C_c = 1, \ D_c = 50/0.15. \tag{3.44}$$

For the controller (3.40), the matrix Q had not been found analytically, however, the global asymptotic stability of the system (3.33), (3.34), and (3.40) has been ascertained using a software package.

3.2 Quality Indicators for Random Reference Tracking in Linear Systems

3.2.1 Scenario

In linear systems with deterministic reference signals, the quality of tracking is characterized, in the time domain, by settling time, percent of overshoot, and so on, or, in the frequency domain, by gain and phase margins. In linear systems with random references, the quality of tracking is often quantified by the standard deviation of the error signal, σ_e. However, as it turns out, this measure is too crude to reveal causes of poor tracking. To justify this statement consider the system of Figure 3.11 with $F_\Omega(s)$ given by

$$F_\Omega(s) = \frac{0.306}{s^3 + s^2 + 0.5s + 0.125} \tag{3.45}$$

(which is a third order Butterworth filter with $\Omega = 0.5$ and the d.c. gain selected so that the standard deviation at its output is 1) and three different $G(s)$:

$$G(s) = \frac{24.8}{(s+5)(s+10)}, \ G(s) = \frac{0.585}{s(s+1.5)}, \ G(s) = \frac{2.31}{s(s+0.02)}. \tag{3.46}$$

In all three cases, the standard deviation of the error signal remains the same, $\sigma_e = 0.67$, while the tracking behavior, shown in Figure 3.12, is qualitatively different. Namely, in Figure 3.12(a) poor tracking appears to be due to static unresponsiveness, in Figure 3.12(b) due to time-lag in comparison with the reference signal, and in Figure 3.12(c) due to oscillatory behavior.

This example shows that σ_e is not capable of discriminating between the causes of poor tracking, and new measures are necessary. Such measures, referred to as *tracking quality indicators*, are introduced in this section.

3.2.2 Random Reference Model

Reference signals considered in this and the subsequent sections are Gaussian colored noise processes obtained by filtering standard white noise by a low-pass filter,

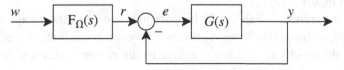

Figure 3.11. Linear tracking system.

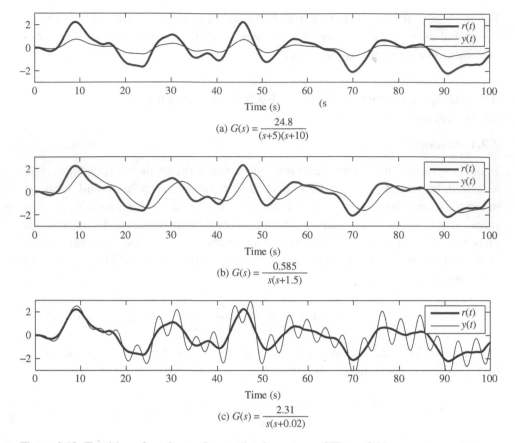

Figure 3.12. Tracking of random reference by the system of Figure 3.11.

$F_\Omega(s)$, with 3dB bandwidth Ω. Although methods developed here apply to any low-pass filter, to be specific we assume that $F_\Omega(s)$ is a third order Butterworth filter with d.c. gain selected so that its output, $r(t)$, has standard deviation equal to 1. In other words, we assume that the reference signal $r(t)$ is the output of

$$F_\Omega(s) = \sqrt{\frac{3}{\Omega}} \left(\frac{\Omega^3}{s^3 + 2\Omega s^2 + 2\Omega^2 s + \Omega^3} \right), \tag{3.47}$$

with standard white noise at its input. Realizations of $r(t)$ for $\Omega = 1$ rad/s and 4 rad/s are shown in Figure 3.13.

Clearly, reference signals could be parameterized in other ways as well, for instance, by higher order Butterworth filters. However, as it is indicated in Subsection 3.2.3 below, the results are not too sensitive to the parameterization involved and, therefore, the one defined by (3.47), is used.

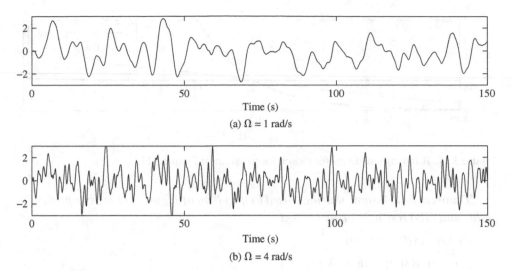

Figure 3.13. Realizations of $r(t)$.

3.2.3 Random Sensitivity Function

In the case of step inputs, the tracking properties of the linear system shown in Figure 3.11 are characterized in the frequency domain by the sensitivity function

$$S(j\omega) = \frac{1}{1 + G(j\omega)}, \quad 0 \le \omega < \infty, \tag{3.48}$$

that is, by the steady state errors in tracking harmonic inputs of different frequencies. Specifically, bandwidth, resonance peak, resonance frequency and d.c. gain of $S(j\omega)$ characterize the quality of step tracking. In this subsection, we introduce the *random sensitivity function* and show that it can be used to define quality indicators for random, rather than deterministic, reference tracking.

Definition 3.2. *The random sensitivity function of the linear feedback system of Figure 3.11 is the standard deviation of the error signal, e, as a function of the bandwidth of the band-limited references $r(t)$:*

$$RS(\Omega) := \sigma_e(\Omega), \quad 0 < \Omega < \infty, \tag{3.49}$$

that is,

$$RS(\Omega) = \|F_\Omega(s)S(s)\|_2, 0 < \Omega < \infty. \tag{3.50}$$

Although $RS(\Omega)$ is just a standard deviation, it is a function of Ω, and, thus, more informative than the single number σ_e obtained for a fixed random input process. For example, the random sensitivity functions of the three systems of Figure 3.12 are qualitatively different although they take the same value for $\Omega = 0.5$ rad/s (see Figure 3.14).

Properties of $RS(\Omega)$ and its relation to $|S(j\omega)|$ are given in the following theorem:

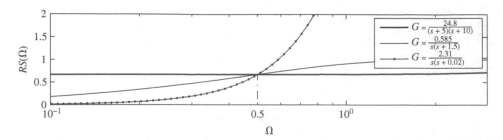

Figure 3.14. Random sensitivity for closed loop system of Figure 3.11.

Theorem 3.4. *Assume that the closed loop system of Figure 3.11 is asymptotically stable and $G(s)$ is strictly proper. Then*

(i) $\lim_{\Omega \to 0} RS(\Omega) = |S(j0)|$;

(ii) $\lim_{\Omega \to \infty} RS(\Omega) = \lim_{\omega \to \infty} |S(j\omega)| = 1$;

(iii) $\sup_{\Omega > 0} RS(\Omega) \leq \sup_{\omega > 0} |S(j\omega)|$.

Proof. *See Section 8.2.*

Figure 3.15 provides comparison of $RS(\Omega)$ and $|S(j\omega)|$ for the system of Figure 3.11 with

$$G(s) = \frac{\omega_n^2}{s(s + 2\zeta\omega_n)} \tag{3.51}$$

for several values of ζ. As indicated in Theorem 3.4, the two curves coincide at zero frequency, converge to 1 as frequencies increase, and $RS(\Omega)$ has a smaller peak. In addition, as one can see, $RS(\Omega)$ has a less pointed peak than $|S(j\omega)|$.

Similar to the usual sensitivity function, $S(j\omega)$, which is characterized by its

$$\text{d.c. gain} = |S(j0)|, \tag{3.52}$$

$$\omega_{BW} = \min\{\omega : \ |S(j\omega)| = 1/\sqrt{2}\}, \tag{3.53}$$

$$\omega_r = \arg\sup_{\omega > 0} |S(j\omega)|, \tag{3.54}$$

$$M_r = \sup_{\omega > 0} |S(j\omega)|, \tag{3.55}$$

the random sensitivity function, $RS(\Omega)$, is also characterized by four quantities:

Random d.c. gain:

$$R_{dc} := \lim_{\Omega \to 0} RS(\Omega). \tag{3.56}$$

Random bandwidth:

$$R\Omega_{BW} := \min\{\Omega : \ RS(\Omega) = 1/\sqrt{2}\}. \tag{3.57}$$

Random resonance frequency:

$$R\Omega_r := \arg\sup_{\Omega > 0} RS(\Omega). \tag{3.58}$$

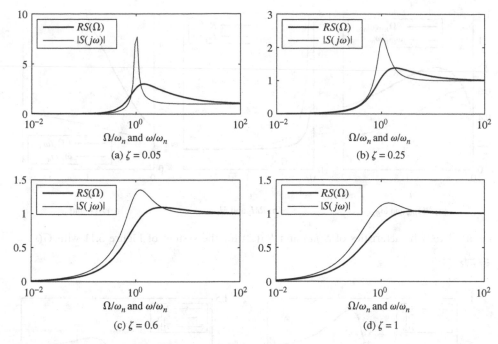

Figure 3.15. Comparison of $RS(\Omega)$ and $|S(j\omega)|$ for the system of Figure 3.11 with $G(s) = \frac{\omega_n^2}{s(s+2\zeta\omega_n)}$.

Random resonance peak:

$$RM_r := \sup_{\Omega > 0} RS(\Omega). \tag{3.59}$$

As stated in Theorem 3.4, the d.c. gains of $RS(\Omega)$ and $S(j\omega)$ are the same, that is,

$$R_{dc} = |S(j0)|.$$

The bandwidths, resonance frequencies, and resonance peaks of $RS(\Omega)$ and $|S(j\omega)|$ exhibit different behavior. They are illustrated in Figure 3.16 for $G(s)$ of (3.51). As one can see, for practical values of ζ, $R\Omega_{BW}$ and $R\Omega_r$ are larger than ω_{BW} and ω_r respectively, while RM_r is smaller than M_r.

Note that selecting higher order Butterworth filters does not significantly change the behavior of $RS(\Omega)$. Indeed Figure 3.17 illustrates random sensitivity functions, calculated according to (3.50), but for Butterworth filters of order 3, 5, and 7, for $G(s)$ given in (3.51). Clearly, $RS(\Omega)$ is robust with respect to reference signal parameterization.

It has been assumed throughout the preceding discussion that $F_\Omega(s)$ is a low-pass filter with

- the Butterworth pole locations;
- the standard deviation of the signal at its output $\sigma = 1$.

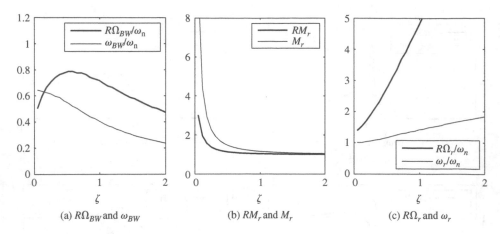

(a) $R\Omega_{BW}$ and ω_{BW} (b) RM_r and M_r (c) $R\Omega_r$ and ω_r

Figure 3.16. Characteristics of $S(j\omega)$ and $RS(\Omega)$ for the system of Figure 3.11 with $G(s) = \frac{\omega_n^2}{s(s+2\zeta\omega_n)}$.

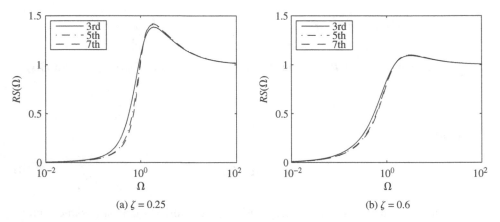

(a) $\zeta = 0.25$ (b) $\zeta = 0.6$

Figure 3.17. $RS(\Omega)$ for Butterworth filters of various orders.

Both of these clauses were used to simplify the presentation. In general, none of them is necessary: Low-pass filters with arbitrary pole-zero locations and σ can be used. The only modification necessary is to normalize σ_e by σ. In other words, rather than (3.50), the random sensitivity function is defined by

$$RS(\Omega) := \frac{||F_\Omega(s)S(s)||_2}{||F_\Omega(s)||_2}, \quad 0 < \Omega < \infty. \tag{3.60}$$

All the conclusions described above remain valid for the definition (3.60) as well.

Computational issues: The random sensitivity function can be calculated as follows: Consider the closed loop system of Figure 3.11, where $F_\Omega(s)$ is the coloring filter and $G(s)$ represents the plant and the controller, that is,

$$G(s) = C(s)P(s). \tag{3.61}$$

Let $S(s)$ be the sensitivity function of this system,

$$S(s) = \frac{1}{1+G(s)}.$$

Introduce the following minimal state space realizations of $S(s)$ and $F_\Omega(s)$:

$$\mathcal{S} = \left[\begin{array}{c|c} A_S & B_S \\ \hline C_S & D_S \end{array} \right]$$

and

$$\mathcal{F}_\Omega = \left[\begin{array}{c|c} A_{\mathcal{F}}(\Omega) & B_{\mathcal{F}}(\Omega) \\ \hline C_{\mathcal{F}} & 0 \end{array} \right],$$

where the representation of $F_\Omega(s)$ is in observable canonical form, so that $C_{\mathcal{F}}$ is independent of Ω. Clearly,

$$\mathcal{S}\mathcal{F}_\Omega = \left[\begin{array}{c|c} A^* & B^* \\ \hline C^* & D^* \end{array} \right],$$

where

$$A^* = \left[\begin{array}{c|c} A_{\mathcal{F}} & 0 \\ \hline B_S C_{\mathcal{F}} & A_S \end{array} \right], B^* = \left[\begin{array}{c} B_{\mathcal{F}} \\ 0 \end{array} \right],$$

$$C^* = \left[\begin{array}{c|c} D_S C_{\mathcal{F}} & C_S \end{array} \right], D^* = 0.$$

Then, for each Ω, the random sensitivity function can be computed as

$$\|F_\Omega S\|_2^2 = \mathrm{tr}\left(C^* W C^{*T} \right), \tag{3.62}$$

where W is the solution of the Lyapunov equation

$$A^* W + W A^{*T} + B^* B^{*T} = 0. \tag{3.63}$$

Solving (3.63) analytically can be accomplished using standard symbolic manipulation software. This provides a means to analytically evaluate the indicators for various closed loop pole locations.

As an example, consider $C(s)P(s)$ given by

$$C(s)P(s) = \frac{\omega_n^2}{s(s+2\zeta\omega_n)}, \tag{3.64}$$

so that the closed loop transfer function is the prototype second order system

$$T(s) = \frac{\omega_n^2}{s^2 + 2\zeta\omega_n s + \omega_n^2}, \tag{3.65}$$

where ω_n and ζ are the natural frequency and damping ratio, respectively. Assume that the coloring filter is a first order system given by

$$F_\Omega(s) = \frac{\Omega}{s+\Omega}. \tag{3.66}$$

Then, using the above approach and noting that

$$||F_\Omega(s)||_2 = \sqrt{\frac{\Omega}{2}},$$

based on (3.60) we obtain:

$$RS\left(\Omega, \omega_n, \zeta\right) = \sqrt{\frac{\Omega\left(2\omega_n\zeta^2 + 2\Omega\zeta + \omega_n\right)}{2\zeta\left(2\Omega\omega_n\zeta + \Omega^2 + \omega_n^2\right)}}. \tag{3.67}$$

3.2.4 Tracking Quality Indicators

Similarly to gain and phase margins, which in most cases characterize the quality of deterministic reference tracking, tracking quality indicators are numbers associated with $RS(\Omega)$, which, in most cases, predict the quality of random reference tracking.

We introduce three indicators. The first two are defined as follows:

$$I_1 := R_{dc}, \tag{3.68}$$

$$I_2 := \frac{\Omega}{R\Omega_{BW}}, \tag{3.69}$$

where, as before, Ω is the bandwidth of the reference signal. The first indicator characterizes the level of static responsiveness, while the second characterizes dynamic properties such as lagging or oscillatory behavior. When both I_1 and I_2 are small, random reference tracking is typically good. For instance, if $G(s) = \frac{2.31}{s(s + 0.02)}$ and $\Omega = 0.05$ rad/sec, the indicators take values

$$I_1 = 0, \quad I_2 = 0.083,$$

and the tracking, as shown in Figure 3.18, is good: signals $r(t)$ and $y(t)$ are practically indistinguishable. When I_1 is large and I_2 is small, tracking is poor due to static unresponsiveness of the closed loop. For instance, in the system of Figure 3.12(a), $I_1 = 0.668$, $I_2 = 0.187$ and the loss of tracking is due to static unresponsiveness.

When I_1 is small and I_2 is large, tracking quality is again poor but due to either lagging or oscillatory behavior. To discriminate between these two cases, we introduce

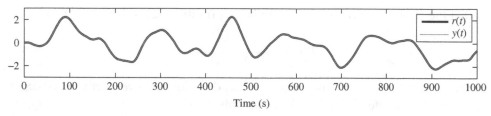

Figure 3.18. Response of the system in Figure 3.11 with $G(s) = \frac{2.31}{s(s + 0.02)}$ and $F_\Omega(s)$ as in (3.47) with $\Omega = 0.05$ rad/sec.

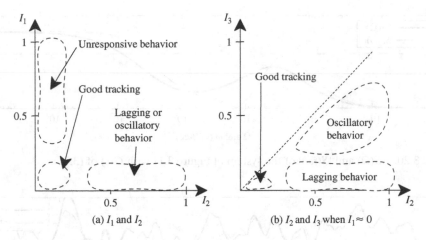

(a) I_1 and I_2 (b) I_2 and I_3 when $I_1 \approx 0$

Figure 3.19. Tracking quality prediction by the indicators.

the third indicator

$$I_3 := \min\left(RM_r - 1, \frac{\Omega}{R\Omega_r} \right). \tag{3.70}$$

This indicator is the minimum of two quantities: the first quantity being large implies the presence of a high resonance peak, while the second being large implies that the resonance is activated by inputs. When I_3 is small, that is, both quantities are small, then oscillatory response does not occur, and the response is of lagging nature; when I_3 is large, the lack of tracking is due to oscillatory behavior. For instance, in the case of Figure 3.12(b) the indicators take values

$$I_1 = 0, \quad I_2 = 0.912, \quad I_3 = 0.0374$$

and the loss of tracking is due to lagging. In the case of Figure 3.12(c) the indicators are

$$I_1 = 0, \quad I_2 = 0.979, \quad I_3 = 0.25$$

and the loss of tracking is due to oscillatory behavior.

Based on these and other judiciously selected examples, the areas of good and bad tracking in the indicator planes (I_1, I_2) and (I_2, I_3) can be represented as shown in Figure 3.19. Exceptions, however, exist: even with small I_1 and I_2, poor tracking is sometimes possible. This happens in the same cases where the usual sensitivity function, $S(j\omega)$, does not predict well the step response. For instance, if

$$G(s) = \frac{361.14(s+10.68)(s+0.2209)(s^2+1.118s+5.546)}{s(s+50)(s+0.01)(s^2+0.1s+196)} \tag{3.71}$$

and $\Omega = 4$, then I_1 and I_2 are small ($I_1 = 0$, $I_2 = 0.005$) but tracking is poor. Figures 3.20–3.22 illustrate $|S(j\omega)|$ and $RS(\Omega)$ and the quality of random and step reference tracking. Note that although $S(j\omega)$ has sufficiently large bandwidth and small resonance peak, the quality of step tracking is also poor.

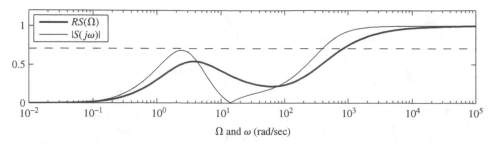

Figure 3.20. $RS(\Omega)$ and $|S(j\omega)|$ of the system of Figure 3.11 with $G(s)$ of (3.71).

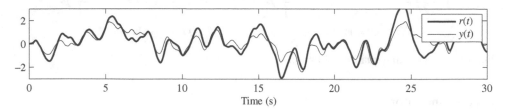

Figure 3.21. Random reference tracking of the system of Figure 3.11 with $G(s)$ of (3.71) and $F_\Omega(s)$ as in (3.47) with $\Omega = 4$ rad/sec.

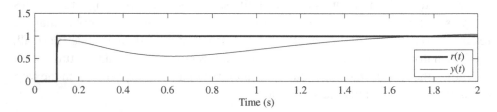

Figure 3.22. Step tracking of the system of Figure 3.11 with $G(s)$ of (3.71).

Tracking quality indicators can be used as specifications for the design of tracking systems. An example of such an application is discussed in the next subsection.

Computational issues: The indicators I_1–I_3 can be computed by using (3.56)–(3.59) and the approach outlined in Subsection 3.2.3 for evaluating the random sensitivity function.

3.2.5 Application: Linear Hard Disk Servo Design

The primary tasks of the hard disk servo controller are track seeking and track following. Track seeking aims at moving a read/write head from one track to another. Track following causes the head to follow the track motion while on track. Tracks to be followed are not perfectly circular due to various sources such as disk surface defects and irregularities of many types, drive vibrations, electrical noise, and so on. These sources are divided into two groups: repeatable runout (RRO) and non-repeatable

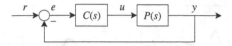

Figure 3.23. Hard disk servo system.

runout (NRRO). Both RRO and NRRO are typically modeled as bandlimited Gaussian processes. Therefore, track following can be viewed as a random reference tracking problem.

The block diagram of the hard disk servo system is shown in Figure 3.23 where $P(s)$ is the head positioning unit, $C(s)$ is the controller to be designed, y is the position of the head in track numbers, and u is the actuator input in volts. The plant $P(s)$ is often modeled as

$$P(s) = \frac{4.3827 \times 10^{10} s + 4.3247 \times 10^{15}}{s^2(s^2 + 1.5962 \times 10^3 s + 9.7631 \times 10^7)}. \tag{3.72}$$

The reference $r(t)$ is assumed to be a zero-mean Gaussian process with variance $\sigma^2_{RRO} + \sigma^2_{NRRO}$, where $\sigma_{RRO} = 0.25$ and $\sigma_{NRRO} = 0.125$. In addition, we assume that the bandwidth of $r(t)$ can be as large as twice the rotation speed of the hard disk ($\omega_0 = 346$ rad/sec), that is, 692 rad/sec.

Next, we introduce specifications for controller design:

1. $I_1 = 0$. This specification is introduced in order to ensure head positioning without any bias.
2. $I_2 \le 0.35\, \forall \Omega < 692$ rad/sec. This specification is introduced in order to ensure fast tracking.
3. $I_3 \le 0.1\, \forall \Omega < 692$ rad/sec. This specification is introduced in order to ensure non-oscillatory behavior.
4. In addition to the above, one more specification must be introduced, which defines the accuracy of the head positioning: It is required that $3\sigma_e$ be less than 5% of the track width. This is ensured by the following specification:

$$RS(\Omega) < 0.0596, \quad \forall \Omega < 692 \text{ rad/sec.}$$

Indeed, using the fact

$$\sigma_e(\Omega) = RS(\Omega)\sigma_r,$$

and $\sigma_r = \sqrt{\sigma^2_{RRO} + \sigma^2_{NRRO}} = 0.2795$, the specification ensures

$$3\sigma_e = 3RS(\Omega)(0.2795) < (3)(0.0596)(0.2795) < 0.05.$$

Using the $\mathcal{H}_\infty$ technique, we design a controller

$$C(s) = \frac{19949211(s^2 + 324.5s + 44630)(s^2 + 1596.2s + 97631000)}{(s + 400)(s^2 + 102800s + 3.06 \times 10^9)(s^2 + 36250s + 2.78 \times 10^9)}, \tag{3.73}$$

which satisfies all design specifications with $I_1 = 0$, $I_2 \le 0.0553$, $\forall \Omega \le 692$, and $RS(\Omega) \le 0.056$, $\forall \Omega \le 692$.

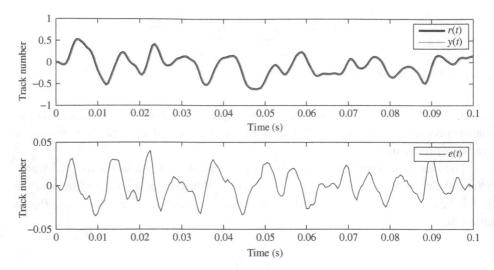

Figure 3.24. Track following performance of the hard disk servo system.

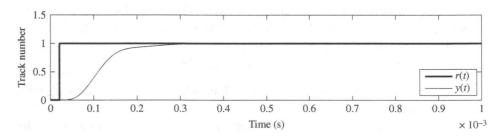

Figure 3.25. Track seeking performance of the hard disk servo system.

Figure 3.24 shows the track following performance for the reference signal of the largest bandwidth, $\Omega = 692$ rad/sec. Obviously, the tracking quality is good and the error is smaller than 5% of the track width.

To illustrate the track seeking performance of controller (3.73), Figure 3.25 shows the unit step response of the closed loop system. As one can see, in this particular case, the overshoot is very close to zero and the settling time is less than 0.3 msec, which is acceptable in most hard disk drives.

3.3 Quality Indicators for Random Reference Tracking in LPNI Systems

3.3.1 Scenario

It has been shown in the previous section that the quality of random reference tracking in linear systems can be characterized by the random sensitivity function and the tracking quality indicators. However, these tools are not applicable to systems with saturating actuators. Indeed, while in the linear case, the random sensitivity is determined by the loop transfer function and, thus, so is the quality of tracking, in systems with saturating actuators, the quality of tracking depends not only on the loop transfer

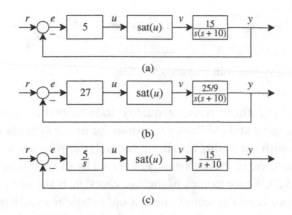

(a)

(b)

(c)

Figure 3.26. Motivating example.

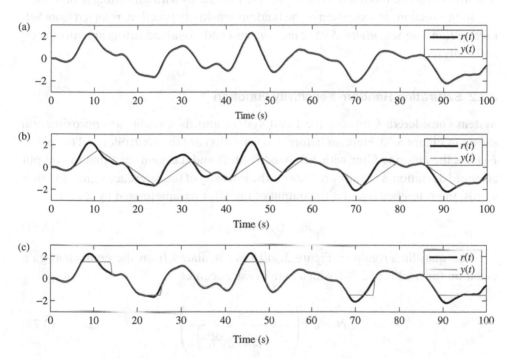

Figure 3.27. Tracking of random references by the systems of Figure 3.26.

function but also on the location of the saturating actuators. For example, the three feedback systems in Figure 3.26 have identical loop transfer function if saturation is ignored, but exhibit qualitatively different tracking behavior for the same reference (see Figure 3.27). Indeed, the system of Figure 3.26(a) results in excellent reference tracking (see Figure 3.27(a)). The system of Figure 3.26(b) exhibits poor tracking, which seems to be due to rate limitation (see Figure 3.27(b)). The system of Figure 3.26(c) shows relatively poor quality of tracking, which seems to be due to amplitude limitation and controller windup. Thus, it is clear that the method developed in the previous section is not directly applicable to the nonlinear case.

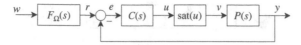

Figure 3.28. Tracking system with saturating actuator.

Furthermore, the three tracking quality indicators that characterize static responsiveness, lagging and oscillatory response for linear systems seem to be deficient for systems with saturating actuators. Indeed, the responses of the system of Figure 3.28 for the same reference signal, generated by the coloring filter (3.45), are shown in Figure 3.29, where in each of the five cases, σ_e is the same ($\sigma_e = 0.67$), but the reasons for poor tracking seem to be not only static responsiveness, lagging or oscillatory behavior. Therefore, the random sensitivity function and tracking quality indicators must be modified to be applicable to systems with saturating actuators.

In this section, an extension of the random sensitivity function, referred to as *Saturating Random Sensitivity (SRS) Function*, and additional indicators are introduced and analyzed.

3.3.2 Saturating Random Sensitivity Function

System considered: Consider the LPNI system and its quasilinear approximation shown in Figure 3.30. Here, as before, $C(s)$ and $P(s)$ are the controller and the plant, $F_\Omega(s)$ is the coloring filter with 3dB bandwidth Ω and d.c. gain such that the output standard deviation is 1, and σ_r represents the intensity of the reference signal. In other words, the reference signal is the output of the filter parameterized by Ω and σ_r:

$$F_{\Omega,\sigma_r}(s) = \sigma_r F_\Omega(s). \tag{3.74}$$

The quasilinear gain in Figure 3.30(b) is calculated from the equivalent gain equation for tracking (2.50), which in this case becomes

$$N = \mathrm{erf}\left(\frac{\alpha}{\sqrt{2}\left\|\frac{\sigma_r F_\Omega(s)C(s)}{1+NP(s)C(s)}\right\|_2}\right). \tag{3.75}$$

For the quasilinear system of Figure 3.30(b), we define below the notion of the *saturating random sensitivity function*.

SRS definition and properties: Using the system of Figure 3.30 we introduce:

Definition 3.3. *The saturating random sensitivity function of the LPNI system of Figure 3.30(a), $SRS(\Omega,\sigma_r)$, is the random sensitivity function of the stochastically linearized system of Figure 3.30(b) normalized by the intensity of the reference signal σ_r:*

$$SRS(\Omega,\sigma_r) := \frac{RS(\Omega)}{\sigma_r}, \ 0 < \Omega < \infty, \ \sigma_r > 0. \tag{3.76}$$

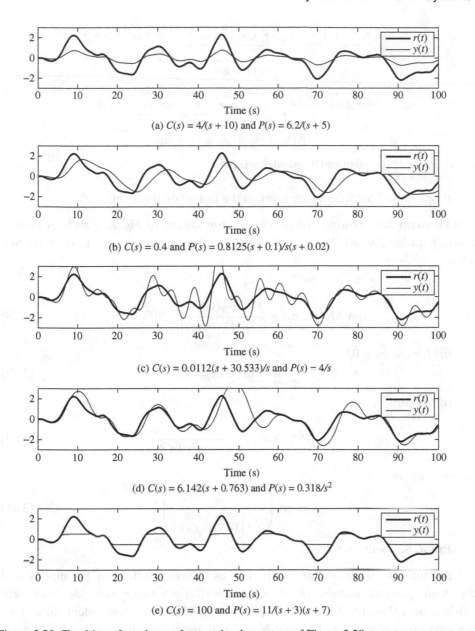

Figure 3.29. Tracking of random reference by the system of Figure 3.28.

where $RS(\Omega)$ is defined by (3.50) if the $||F_\Omega(s)||_2 = 1$ or by (3.60) if the $||F_\Omega(s)||_2 \neq 1$.

Clearly, this function can be calculated as

$$SRS(\Omega, \sigma_r) = \left\| \frac{F_\Omega(s)}{1 + NP(s)C(s)} \right\|_2 / ||F_\Omega(s)||_2, \qquad (3.77)$$

where N is a solution of (3.75), which is assumed, for simplicity, to be unique.

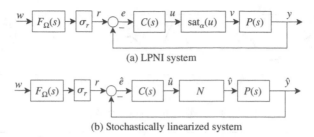

(a) LPNI system

(b) Stochastically linearized system

Figure 3.30. LPNI system and its quasilinearization.

Properties of $SRS(\Omega, \sigma_r)$ are given in the following theorem:

Theorem 3.5. *Assume that the closed loop system of Figure 3.30(b) is asymptotically stable for all $N \in (0, 1]$, $P(s)$ is strictly proper and $C(s)$ is proper. Then,*
(i) for any $\Omega > 0$,

$$\lim_{\sigma_r \to 0} SRS(\Omega, \sigma_r) = \left\| \frac{F_\Omega(s)}{1 + P(s)C(s)} \right\|_2 / \|F_\Omega(s)\|_2; \qquad (3.78)$$

(ii) for any $\sigma_r > 0$,

$$\lim_{\Omega \to \infty} SRS(\Omega, \sigma_r) = 1; \qquad (3.79)$$

(iii) for any $\sigma_r > 0$,

$$\lim_{\Omega \to 0} SRS(\Omega, \sigma_r) = \left| \frac{1}{1 + NP(0)C(0)} \right|, \qquad (3.80)$$

where N satisfies

$$N = \mathrm{erf}\left(\frac{\alpha}{\sqrt{2} \left| \frac{\sigma_r F_\Omega(0)C(0)}{1 + NP(0)C(0)} \right|} \right). \qquad (3.81)$$

Proof. See Section 8.2

An interpretation of this theorem is as follows: Statement (i) implies that for small reference signals, $SRS(\Omega, \sigma_r)$ and $RS(\Omega)$ practically coincide. Statement (ii) indicates that, for large Ω, $SRS(\Omega, \sigma_r)$ and $RS(\Omega)$ are again identical and no tracking takes place. Finally, since $N \leq 1$, statement (iii) shows that $SRS(\Omega, \sigma_r)$ is typically larger than $RS(\Omega)$ (at least for low frequencies) and, therefore, the presence of saturating actuators impedes tracking.

Illustrations of $SRS(\Omega, \sigma_r)$ for

$$F_\Omega(s) = \sqrt{\frac{3}{\Omega}} \left(\frac{\Omega^3}{s^3 + 2\Omega s^2 + 2\Omega^2 s + \Omega^3} \right) \qquad (3.82)$$

are given in Figure 3.31. Specifically, Figure 3.31(a)–(c) shows $SRS(\Omega, \sigma_r)$ for each system of Figure 3.26, while Figure 3.31(d)–(h) shows $SRS(\Omega, \sigma_r)$ for the systems of Figure 3.29.

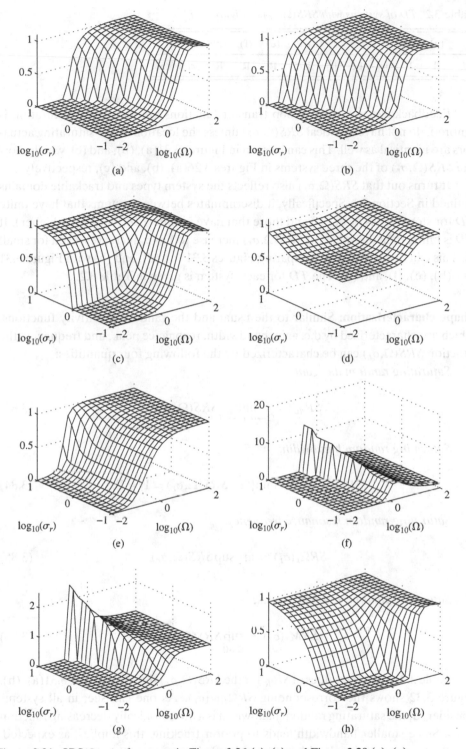

Figure 3.31. $SRS(\Omega, \sigma_r)$ of systems in Figure 3.26 (a)–(c) and Figure 3.29 (a)–(e).

Table 3.2. *TD of systems with SRS(Ω, σ_r) of Figure 3.31*

	(a)	(b)	(c)	(d)	(e)	(f)	(g)	(h)
TD	**R**	**R**	$|r| < 1.5$	$|r| < 3.75$	**R**	**R**	**R**	$|r| < 0.53$

Feedback systems, whose loop transfer functions are identical if saturation is ignored, do not have identical $SRS(\Omega, \sigma_r)$ unless the location of the saturating actuators are identical as well. This can be seen in Figures 3.31(a), (b), and (c), which show the $SRS(\Omega, \sigma_r)$ of the three systems in Figures 3.26(a), (b), and (c), respectively.

It turns out that $SRS(\Omega, \sigma_r)$ also reflects the system types and trackable domains defined in Section 3.1. Specifically, it discriminates between systems that have finite TD for steps (types 0_S or 0_S^+) and those that have infinite TD (types 1_S or higher). If TD is finite, then the values of $SRS(\Omega, \sigma_r)$ increase to 1 as σ_r increases, even for small Ω. This can be observed by comparing Figures 3.31 (c), (d), and (h) with Figure 3.31 (a), (b), (e), (f), and (g). The TD for each system is given in Table 3.2.

Shape characterization: Similar to the usual and the random sensitivity functions, which are characterized by d.c. gain, bandwidth, resonance peak, and frequency, the function $SRS(\Omega, \sigma_r)$ can be characterized by the following four quantities:
Saturating random d.c. gain:

$$SR_{dc} := \lim_{\Omega \to 0, \, \sigma_r \to 0} SRS(\Omega, \sigma_r). \tag{3.83}$$

Saturating random bandwidth:

$$SR\Omega_{BW}(\sigma_r) := \min\{\Omega : \ SRS(\Omega, \sigma_r) = 1/\sqrt{2}\}. \tag{3.84}$$

Saturating random resonance frequency:

$$SR\Omega_r(\sigma_r) := \arg\sup_{\Omega > 0} SRS(\Omega, \sigma_r). \tag{3.85}$$

Saturating random resonance peak:

$$SRM_r(\sigma_r) := \sup_{\Omega > 0} SRS(\Omega, \sigma_r). \tag{3.86}$$

Table 3.3 shows the values of SR_{dc} for the $SRS(\Omega, \sigma_r)$ shown in Figure 3.31(a)–(h). Figure 3.32 shows the corresponding $SR\Omega_{BW}(\sigma_r)$. As one can see, in all systems considered, the saturating random bandwidth is a monotonically decreasing function of σ_r. Since smaller bandwidth leads to poorer tracking, this implies, as expected, that the tracking of signals with large amplitude in systems with saturating actuators is generically poor. An interesting effect can be observed in curves (c), (d), and (h) of Figure 3.32, where the saturating random bandwidth drops almost in a relay manner

Table 3.3. SR_{dc} of systems with $SRS(\Omega, \sigma_r)$ of
Figure 3.31

	(a)	(b)	(c)	(d)	(e)	(f)	(g)	(h)
SR_{dc}	0	0	0	0.67	0	0	0	0.019

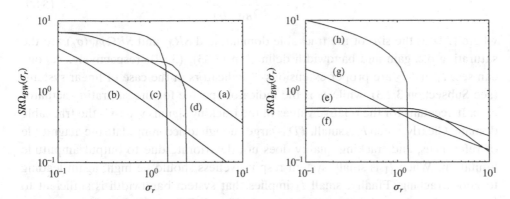

Figure 3.32. Saturating random bandwidth for $SRS(\Omega, \sigma_r)$ of Figure 3.31.

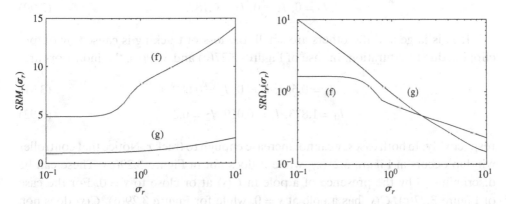

Figure 3.33. Saturating random resonance peak and frequency for $SRS(\Omega, \sigma_r)$ of Figure 3.31(f) and (g).

to zero at certain value of σ_r (practically infinite roll-off rate). This phenomenon is caused by finite TD of the systems, and the loss of tracking is due to "clipping" of the output.

The behavior of the saturating random resonance peak and the saturating random resonance frequency as a function of σ_r are shown in Figure 3.33. Clearly, the former is increasing while the latter is decreasing, both leading to poorer tracking.

3.3.3 Tracking Quality Indicators

We introduce three indicators defined as follows:

$$I_0 := \frac{\sigma_r}{|TD|}, \tag{3.87}$$

$$I_1 := SR_{dc}, \tag{3.88}$$

$$I_2 := \frac{\Omega}{SR\Omega_{BW}(\sigma_r)}, \tag{3.89}$$

where $|TD|$ is the size of the trackable domain, and SR_{dc} and $SR\Omega_{BW}(\sigma_r)$ are the saturating d.c. gain and bandwidth defined in (3.83), (3.84), respectively. As one can see, I_1 and I_2 are proper extensions of indicators in the case of linear systems (see Subsection 3.2.3), while I_0 is the indicator specific for the saturating actuator case: It accounts for the relative values of the tracking signals vis-a-vis the trackable domain. Clearly, when I_0 is small, TD is large enough to accommodate the magnitude of references, and tracking quality does not deteriorate due to output amplitude limitation. When I_1 is small, system responsiveness should be high, again leading to good tracking. Finally, small I_2 implies that system bandwidth is sufficient to accommodate the variability of the reference signal. Thus, if all three indicators are small, the quality of tracking, in most cases, is good. This is illustrated by the example of Figure 3.27(a), for which

$$I_0 = 0, \ I_1 = 0, \ I_2 = 0.187. \tag{3.90}$$

If I_0 is large and the others are small, the loss of tracking is caused by output clipping due to saturation. In case of Figures 3.27(c) and 3.29(e), the indicators are

$$I_0 = 0.667, \ I_1 = 0, \ I_2 = 0.089, \tag{3.91}$$

$$I_0 = 1.873, \ I_1 = 0.019, \ I_2 = 0.2, \tag{3.92}$$

respectively. In both cases, y cannot increase enough to track r. Notice that controller windup occurs in Figure 3.27(c), while it does not in Figure 3.29(e). These can be discriminated by the presence of a pole in $C(s)$ at or close to $s = 0$. For the case of Figure 3.27(c), $C(s)$ has a pole at $s = 0$, while for Figure 3.29(e), $C(s)$ does not. Hence, the presence of a pole of $C(s)$ at or close to $s = 0$ serves to predict controller windup behavior.

If I_1 is large and the others are small, the loss of tracking is due to static unresponsiveness of the output as illustrated in Figure 3.29(a), for which

$$I_0 = 0.267, \ I_1 = 0.668, \ I_2 = 0.187. \tag{3.93}$$

The loss of tracking in this case is due to the linear part of the system rather than actuator saturation.

While I_0 and I_1 characterize the quality of static tracking, I_2 characterizes dynamic properties of the output. If I_2 is large and the others are small, the loss of tracking is due to lagging or oscillatory behavior of the output. To discriminate between the two cases, we use the fourth indicator

$$I_3 := \min \left(SRM_r(\sigma_r) - 1, \frac{\Omega}{SR\Omega_r(\sigma_r)} \right). \qquad (3.94)$$

Again, this indicator is a proper extension of indicator I_3 in the linear case (see Subsection 3.2.3). If I_3 is small, then the output is of lagging nature, otherwise it is oscillatory in its behavior. The responses in Figures 3.29(b) and 3.27(b) have indicators

$$I_0 = 0, \ I_1 = 0, \ I_2 = 0.791, \ I_3 = 0.003, \qquad (3.95)$$

$$I_0 = 0, \ I_1 = 0, \ I_2 = 0.807, \ I_3 = 0, \qquad (3.96)$$

respectively, and the responses are of lagging nature. However, the cause of lagging seems to be different in each case. In Figure 3.29(b), it is due to the linear part of the system, whereas in Figure 3.27(b), it is due to the saturating actuator. The two cases can be differentiated by the value of the equivalent gain N in (3.75) since it characterizes the extent of saturation. If N is close to 1, then the lagging is caused by the linear part, otherwise it is due to actuator saturation. In the case of Figure 3.27(b), $N = 0.049$, whereas in the case of Figure 3.29(b), $N = 1$. Therefore, the value of the equivalent gain N allows to distinguish whether the poor dynamic tracking is caused by the linear part or the saturation.

Figures 3.29(c) and 3.29(d) illustrate the case where I_3 is large. The indicators are

$$I_0 = 0, \ I_1 = 0, \ I_2 = 0.977, \ I_3 = 0.756, \qquad (3.97)$$

$$I_0 = 0, \ I_1 = 0, \ I_2 = 1, \ I_3 = 0.252, \qquad (3.98)$$

respectively, and the poor tracking is due to oscillatory response. Again the cause of oscillation in each case is different and can be differentiated by N. In the case of Figure 3.29(c), $N = 1$, and the oscillatory behavior is caused by the linear part of the system. In the case of Figure 3.29(d), $N = 0.192$, and the oscillations are caused by the saturating actuators. In this latter case, the oscillations are associated more with stability of the system rather than resonance of the feedback loop.

If more than one indicator is large, the tracking is poor due to more than one reason. For instance, the response of the system in Figure 3.26(c) with input standard deviation increased by 50% is shown in Figure 3.34. The indicators are

$$I_0 = 1, \ I_1 = 0, \ I_2 = 4.705, \ I_3 = 0.078, \qquad (3.99)$$

and the poor tracking is due to both amplitude limitation and lagging of y.

As described above, the first three indicators I_0, I_1, and I_2 should be examined first, and if at least one of them is large, then indicator I_3, poles of $C(s)$, and the value of N should be considered in order to further identify the causes of poor tracking. The diagnostic flowchart, which implements this strategy, is shown in Figure 3.35.

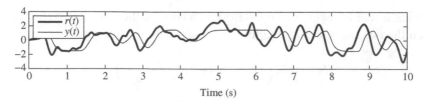

Figure 3.34. The response of the system in Figure 3.26(c) with $\sigma_r = 1.5$ and $\Omega = 10$.

Table 3.4. *Tracking quality indicators for responses shown in Figures 3.27, 3.29, and 3.34*

Figure	I_0	I_1	I_2	I_3	N	$C(s)$ has pole at or close to $s = 0$?	Quality of tracking as determined by the diagnostic flow chart in Figure 3.35
3.27(a)	0	0	0.19	0.02			good
3.27(b)	0	0	0.81	0.01	0.05		lagging (nonlinear)
3.27(c)	0.67	0	0.09	0.02		Yes	small TD, windup
3.29(a)	0.27	0.67	0.19	0.01			static unresponsiveness
3.29(b)	0	0	0.79	0	1		lagging (linear)
3.29(c)	0	0	0.98	0.76	1	Yes	oscillatory (linear)
3.29(d)	0	0	1.00	0.25	0.19		oscillatory (nonlinear)
3.29(e)	1.87	0.02	0.20	0.00			small TD
3.34	1	0	4.71	0.08	0.47	Yes	small TD, lagging, windup

According to our experience, each indicator is considered to be large if

$$I_0 > 0.4, \quad I_1 > 0.1, \quad I_2 > 0.4, \quad I_3 > 0.2, \tag{3.100}$$

respectively. Table 3.4 summarizes the values of tracking quality indicators for all systems considered in this chapter.

Note that as the level of saturation α tends to infinity, $|TD|$ approaches infinity and N approaches one. In this case, the situation becomes similar to the case studied in Section 3.2, and indicators I_1, I_2, and I_3 are sufficient to predict the quality of tracking. Furthermore, the dependency of indicators I_2 and I_3 on σ_r decreases, and they reduce to those defined for linear systems in Section 3.2. In this sense, they are proper extensions of the indicators defined in Section 3.2.

The indicator approach is effective in most cases, however, as explained in Section 3.2, in some cases, it may not be effective. In such cases, the complete shape of $SRS(\Omega, \sigma_r)$ should be investigated to predict the quality of tracking.

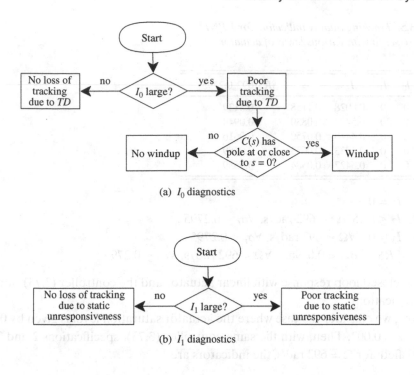

(a) I_0 diagnostics

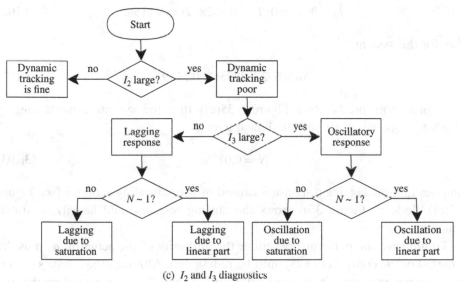

(b) I_1 diagnostics

(c) I_2 and I_3 diagnostics

Figure 3.35. Diagnostic flowchart for quality of tracking in systems with saturating actuators.

3.3.4 Application: LPNI Hard Disk Servo Design

In this subsection, the hard disk servo example of Subsection 3.2.4 is revisited with actuator saturation constraint. The plant is given by (3.72) and the design specifications remain as in the linear case (with the addition of σ_r and with $SRS(\Omega, \sigma_r)$ used instead of $RS(\Omega)$), that is,

Table 3.5. *Tracking quality indicators for LPNI*
hard disc servo with various levels of actuator
authority

α	I_0	I_1	I_2	I_3	$SRS(692, 0.2795)$
0.002	0	0	0.8028	0.1158	0.5139
0.004	0	0	0.5497	0.0830	0.0944
0.006	0	0	0.4440	0.0750	0.0616
0.008	0	0	0.3833	0.0633	0.0569
0.01	0	0	0.3427	0.0584	0.0559

1. $I_1 = 0$
2. $I_2 \leq 0.35 \; \forall \Omega < 692$ rad/s, $\forall \sigma_r \leq 0.2795$
3. $I_3 \leq 0.1 \; \forall \Omega < 692$ rad/s, $\forall \sigma_r \leq 0.2795$
4. $SRS(\Omega, \sigma_r) < 0.0596, \quad \forall \Omega < 692$ rad/s, $\forall \sigma_r \leq 0.2795$

The closed loop response with linear actuator and the controller (3.73) satisfied the specifications.

Now, we consider the case where the actuator saturation level is given by 0.002, that is, $\alpha = 0.002$. Then, with the same controller (3.73), specifications 2 and 3 are not satisfied: for $\Omega = 692$ rad/s, the indicators are

$$I_0 = 0, \; I_1 = 0, \; I_2 = 0.8028, \; I_3 = 0.1158. \tag{3.101}$$

Also, for this system,

$$SRS(692, 0.2795) = 0.5139. \tag{3.102}$$

These indicators predict (see Figure 3.35(c)) that the system exhibits lagging behavior. Moreover, since the quasilinear gain is

$$N = 0.0175, \tag{3.103}$$

it implies that the lagging response is caused by the saturating actuator (see Figure 3.35(c)). Indeed, Figure 3.36 shows the lagging response and heavily saturated actuation.

To improve the performance, either the authority of the actuator, α, must be increased or the controller (3.73) must be redesigned. Although these issues of performance recovery are discussed in details in Chapter 7, below we follow the first of the above two avenues and select α for which performance specifications 1–4 are met.

Table 3.5 presents the tracking quality indicators for various values of α in the LPNI system with plant (3.72) and controller (3.73). Clearly, increasing α leads to improved performance, and with $\alpha = 0.01$ all specifications are satisfied. The corresponding behavior in the time domain is illustrated in Figures 3.37–3.40. Thus, to ensure the desired performance with controller (3.73), $\alpha = 0.01$ (or at least 0.008) should be used.

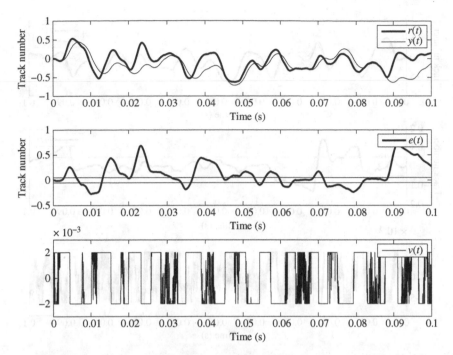

Figure 3.36. Tracking in the hard disk servo with actuator saturation $\alpha = 0.002$.

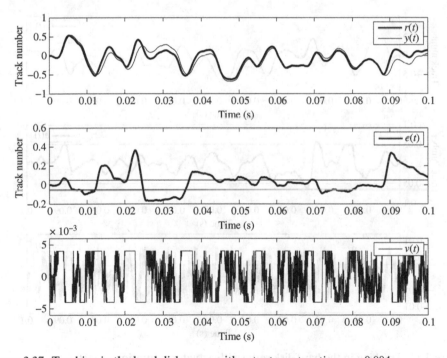

Figure 3.37. Tracking in the hard disk servo with actuator saturation $\alpha = 0.004$.

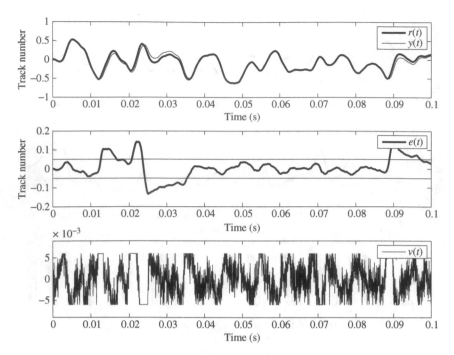

Figure 3.38. Tracking in the hard disk servo with actuator saturation $\alpha = 0.006$.

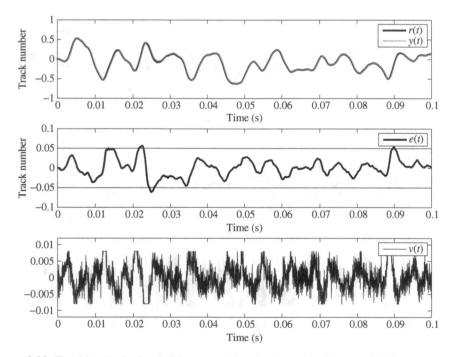

Figure 3.39. Tracking in the hard disk servo with actuator saturation $\alpha = 0.008$.

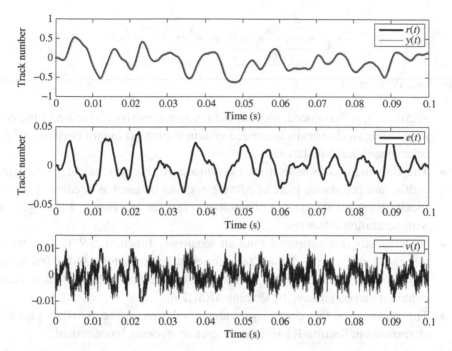

Figure 3.40. Tracking in the hard disk servo with actuator saturation $\alpha = 0.01$.

3.4 Summary

- Unlike the linear case, the type of LPNI systems with saturating actuators is determined by the plant (rather than loop gain) poles at the origin. The controller poles at the origin also play a role but only from the point of view of steady state errors.
- The LPNI systems with saturating actuators have a finite trackable domain for step inputs, unless the plant has a pole at the origin. Controller poles at the origin decrease (rather than increase) the trackable domain.
- The variance of the error signal is a poor predictor of the quality of random reference tracking: for the same error signal variance, the quality of tracking and reasons for the track loss may be qualitatively different.
- To quantify the quality of random reference tracking in linear systems, the so-called random sensitivity function, $RS(\Omega)$, is introduced, where Ω is the bandwidth of the reference signal. This function plays the same role in tracking random references as the usual sensitivity function, $S(j\omega)$, in tracking harmonic references.
- Using d.c. gain, bandwidth, and resonance peak/frequency of $RS(\Omega)$, tracking quality indicators are introduced, which play the same role for predicting the quality of random signals tracking as the gain and phase margins of $S(j\omega)$ do for predicting the quality of deterministic reference tracking.
- Similar developments are carried out for LPNI systems with saturating actuators. In this case, the so-called saturating random sensitivity function,

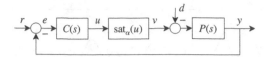

Figure 3.41. Problem 3.4.

SRS(Ω, σ_r), is introduced, which is the random sensitivity function of the corresponding stochastically linearized system with the standard deviation of the reference signal equal to σ_r.

- The tracking quality indicators, introduced on the basis of d.c. gain, bandwidth, and resonance peak of $SRS(\Omega, \sigma_r)$, can be used as predictors of and specification for the quality of random reference tracking in LPNI systems with saturating actuators.
- Ideologically, the saturated random sensitivity function, $SRS(\Omega, \sigma_r)$, transfers the frequency (ω) domain techniques of linear systems to the frequency (Ω) domain techniques of LPNI systems: the desired LPNI system tracking behavior can be ensured by shaping $SRS(\Omega, \sigma_r)$.
- The material of this chapter provides a solution of the Analysis problem introduced in Section 1.2 as far as reference tracking is concerned.

3.5 Problems

Problem 3.1. Determine the type of the system shown in Figure 3.3 with the $P(s)$ and $C(s)$ given below:

(a) $P(s) = \dfrac{10}{s + 0.1}$, $C(s) = 10$;

(b) $P(s) = \dfrac{10}{s + 0.2}$, $C(s) = \dfrac{10s + 3}{s}$;

(c) $P(s) = \dfrac{90}{s}$, $C(s) = 1$;

(d) $P(s) = \dfrac{1}{s(s + 2)}$, $C(s) = \dfrac{5s + 1}{s}$;

(e) $P(s) = \dfrac{0.1}{s^2}$, $C(s) = 3s + 1$;

(f) $P(s) = \dfrac{s + 1}{s^2}$, $C(s) = 5$.

Problem 3.2. Determine the step, ramp, and parabolic trackable domains for the systems of Problem 3.1 with $\alpha = 1$, $\alpha = 2$, and $\alpha = 5$.

Problem 3.3. Consider the system of Figure 3.3 with $\alpha = 1$. For all systems defined in Problem 3.1, determine the errors in tracking step, ramp, and parabolic reference signals under the assumption that r_0, r_1, and r_2 belong to their respective trackable domains.

Problem 3.4. Similar to the trackable domain, the notion of *rejectable domain* can be introduced. To accomplish this, consider the LPNI system of Figure 3.41 and assume that the disturbance, d, is a constant.

(a) Analyze the steady state value of the error signal e as a function of d for

(i) $P(s) = \dfrac{1}{s}$, $C(s) = 10$, $\alpha > 0$, and $r = 0$;

(ii) $P(s) = \dfrac{1}{s}$, $C(s) = \dfrac{10s + 1}{s}$, $\alpha > 0$, and $r = 0$.

(b) Based on this analysis, introduce the notion of rejectable domain and quantify it in terms of the plant and controller parameters.

Problem 3.5. Consider again the system of Figure 3.41 and assume that $r \neq 0$. For

$$P(s) = \frac{1}{s}, \quad C(s) = \frac{10s + 1}{s}, \quad \alpha > 0, \tag{3.104}$$

find the ranges of constant r and d for which the steady state value of e is 0.

Problem 3.6. Consider the system of Figure 3.3 and assume that $\alpha = 2$, and the plant is the d.c. motor with the transfer function

$$P(s) = \frac{1}{s(s + 1)}. \tag{3.105}$$

(a) With this motor, design a controller so that the resulting servo tracks ramps and steps with zero steady state error and evaluate the resulting trackable domains.

(b) With the controller designed in part (a), simulate the system and plot the step response for $r_0 = 1$. If necessary, modify the controller to improve the transient response.

(c) With the controller designed in part (b), simulate the system and plot the step response for $r_0 = 50$. Suggest further modifications of the controller, if necessary, to improve the transient response.

Problem 3.7. Consider the reference tracking LPNI system of Figure 3.3 with a saturating actuator and the plant

$$P(s) = \frac{3}{s + 1}. \tag{3.106}$$

(a) What is the smallest system type, which is necessary to ensure step tracking with zero steady state error?

(b) Design a controller and select the smallest level of actuator saturation, which results in zero steady state error with respect to step inputs having heights up to $r_0 = 5$.

(c) Can the LPNI system with the above plant, controller, and a saturating actuator track a ramp input?

Problem 3.8. Consider the LPNI closed loop system of Figure 3.42.

(a) Assume that the actuator and sensor are given by $f(u) = u$ and $g(y) = \text{sat}_\beta(y)$, respectively. Under the assumption that a unique steady state exists, derive

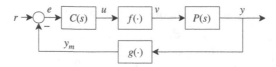

Figure 3.42. Problems 3.8 and 3.9.

expressions for the steady state errors and trackable domains with respect to step and ramp reference signals.

(b) Repeat the same problem for $f(u) = dz_\Delta(u)$ and $g(y) = y$, where $dz_\Delta(\cdot)$ is defined by (2.23).

(c) Repeat the same problem for $f(u) = sat_\alpha(dz_\Delta(u))$ and $g(y) = y$, where $sat_\alpha(dz_\Delta(\cdot))$ is defined by (2.26).

Problem 3.9. Consider the LPNI closed loop system of Figure 3.42 with $f(u) = sat_\alpha(u)$ and $g(y) = y$. Assume that $r(t)$ is a step of size r_0. Make any assumption you wish about the plant and the controller and derive a sufficient condition under which the system has multiple equilibria.

Problem 3.10. Plot the random sensitivity function $RS(\Omega)$ and calculate $R\Omega_{BW}$, $R\Omega_r$, and RM_r of the system shown in Figure 3.11 with $F_\Omega(s)$ in (3.47) and $G(s)$ given below:

(a) $G(s) = \dfrac{5}{s(1+0.5s)(1+0.1s)}$;

(b) $G(s) = \dfrac{10}{s(1+0.5s)(1+0.1s)}$;

(c) $G(s) = \dfrac{500}{(s+1.2)(s+4)(s+10)}$;

(d) $G(s) = \dfrac{10(s+1)}{s(s+2)(s+10)}$;

(e) $G(s) = \dfrac{0.5}{s(s^2+s+1)}$;

(f) $G(s) = \dfrac{10(s+5)}{s(s^2+5s+5)}$.

Problem 3.11. Consider the linear system of Figure 3.11 and assume that

$$G(s) = \frac{K}{s}, \quad F_\Omega(s) = \frac{\sqrt{2\Omega}}{s+\Omega}. \tag{3.107}$$

(a) Derive an analytical expression for the random sensitivity function $RS(\Omega)$.

(b) Find the random bandwidth, $R\Omega_{BW}$, as a function of K; interpret and rationalize the behavior of $R\Omega_{BW}$ as K increases.

(c) Find the random resonance peak, RM_r, as a function of K.

(d) Find the tracking quality indicators I_1, I_2, and I_3, as a function of Ω.

(e) For $K = 1$ and $\Omega = 2$ rad/s, predict the quality of random reference tracking in this system.

(f) For $K = 1$ and $\Omega = 2$ rad/s, simulate the system and compare the results with your prediction.

Problem 3.12. Consider the system in Problem 3.11 with

$$F_\Omega(s) = \frac{\Omega}{s+\Omega}. \tag{3.108}$$

(a) Derive an analytical expression for the random sensitivity function $RS(\Omega)$.
(b) Compare $RS(\Omega)$ with that of Problem 3.11 part (a).
(c) For $\Omega = 10$ rad/s, select K so that the quality of tracking is good (i.e., I_1, I_2, and I_3 are small enough).
(d) Verify by simulations that the quality of tracking is as predicted in part (c).

Problem 3.13. Consider the systems of Problem 3.10.

(a) Compute I_1, I_2, and I_3, for each of the systems as functions of Ω.
(b) For $\Omega = 2$ rad/s, predict the quality of random reference tracking.
(c) Simulate the system with $\Omega = 2$ rad/s and compare the results with your prediction.

Problem 3.14. Consider a d.c. motor with the transfer function

$$P(s) = \frac{1}{s(s+1)} \tag{3.109}$$

and the reference signal generated by filtering standard white noise by

$$F_\Omega(s) = \frac{\Omega}{s+\Omega}, \tag{3.110}$$

and $\Omega = 10$ rad/s.

(a) Design a P-controller for this motor so that the resulting linear servo system tracks the reference defined above with the following tracking quality indicators:

$$I_1 = 0, \quad I_2 = 0.1. \tag{3.111}$$

(b) Simulate this system and verify that the quality of tracking is indeed as expected.
(c) Using simulations, investigate how this system tracks steps.
(d) Using simulations, investigate how the system tracks "slanted steps" defined by

$$u_\Delta(t) = \begin{cases} 0, & t < 0, \\ t/\Delta, & 0 \le t \le \Delta, \\ 1, & t > \Delta. \end{cases} \tag{3.112}$$

Specifically, determine the smallest Δ for which tracking is good and investigate how this Δ relates to Ω.

(e) Assume now that the motor is, in fact, described by

$$P(s) = \frac{110}{s(s+1)(s+10)(s+11)}. \tag{3.113}$$

Simulate the system with the controller designed in part (a) and determine if the idea of dominant poles works for random reference tracking as well.

Problem 3.15. Consider the system of Figure 3.30 with

$$P(s) = \frac{9}{s+0.13}, \quad C(s) = \frac{s+2}{s}, \quad \alpha = 0.2, \tag{3.114}$$

and $F_\Omega(s)$ as given in (3.82).

(a) Plot the saturating random sensitivity function $SRS(\Omega, \sigma_r)$.
(b) For $\sigma_r = 3$ and $\Omega = 1$ rad/s, evaluate I_0, I_1, I_2, and I_3 and using the diagnostic flowcharts, predict the system behavior.
(c) Simulate this system and compare the results with your prediction.

Problem 3.16. Consider the system of Figure 3.30 with

$$P(s) = \frac{1}{s(s+0.3)}, \quad C(s) = \frac{s+0.2}{s+2}, \quad \alpha = 0.1, \tag{3.115}$$

and $F_\Omega(s)$ as given in (3.82).

(a) Plot the saturating random sensitivity function $SRS(\Omega, \sigma_r)$.
(b) For $\sigma_r = 10$ and $\Omega = 0.01$ rad/s, evaluate I_0, I_1, I_2, and I_3 and using the diagnostic flowcharts, predict the system behavior.
(c) Simulate this system and compare the results with your prediction.

Problem 3.17. Consider the LPNI random reference tracking system in Figure 3.30 with

$$P(s) = \frac{1}{s}, \quad C(s) = K, \quad \alpha = 2, \quad F_\Omega(s) = \frac{\sqrt{2\Omega}}{s+\Omega}. \tag{3.116}$$

(a) Plot the saturating random sensitivity function $SRS(\Omega, \sigma_r)$ for $K = 1, 2, 5$, and 10.
(b) For $\sigma_r = 1$, plot the random bandwidth, $SR\Omega_{BW}$, as a function of K; interpret and rationalize the behavior of $R\Omega_{BW}$ as K increases. Compare the results with Problem 3.11 part (b).
(c) For $K = 1, 2, 5$, and 10, plot the saturating random bandwidth, $SR\Omega_{BW}$, as a function of σ_r; interpret and rationalize the behavior of $SR\Omega_{BW}$ as K increases.
(d) Assume that $K = 5$ and $\sigma_r = 2$. Compute the indicators I_0, I_1, I_2, and I_3 for $\Omega = 4$ rad/s and predict tracking quality.
(e) Verify by simulations that the quality of tracking is as predicted in part (d).

Problem 3.18. Consider the LPNI random reference tracking system in Figure 3.30 with

$$P(s) = \frac{1}{s+1}, \quad C(s) = \frac{K(s+2)}{s}, \quad \alpha = 2, \quad F_\Omega(s) = \frac{\sqrt{2\Omega}}{s+\Omega}. \tag{3.117}$$

(a) Plot the saturating random sensitivity function $SRS(\Omega, \sigma_r)$ for $K = 1, 2, 5,$ and 10. Compare the results with Problem 3.17 part (a).

(b) For $K = 1, 2, 5,$ and 10, plot the random bandwidth, $SR\Omega_{BW}$, as a function of σ_r; interpret and rationalize the results using the notion of trackable domain. Compare the results with Problem 3.17 part (c).

Problem 3.19. Consider the servo system designed in part (a) of the Problem 3.14, but assume that the actuator is saturating with $\alpha = 0.1$.

(a) By evaluating $I_0, I_1, I_2,$ and I_3 and using the diagnostic flowcharts, predict the system behavior and draw a likely system trajectory.

(b) Simulate this system and compare with your prediction.

(c) If necessary, redesign the system by selecting a different α so that the behavior is acceptable.

(d) Using simulations, investigate how this system tracks steps with α selected in part (c).

(e) Determine the smallest $\Delta > 0$ for which the tracking of the "slanted" unit step is good, where the slanted step is defined in (3.112).

Problem 3.20. Consider the system of Figure 3.30 with

$$P(s) = \frac{10 - s}{s(s+10)}, \quad \alpha = 3, \tag{3.118}$$

and $F_\Omega(s)$ as given in (3.82).

(a) Design a controller such that the resulting servo satisfies

$$I_0 \leq 0.1, \quad I_1 \leq 0.1, \quad I_2 \leq 0.1, \quad I_3 \leq 0.2, \tag{3.119}$$

for all $\sigma_r \leq 10$, and $\Omega \leq 1$ rad/s.

(b) Simulate this system for $\sigma_r = 10$ and $\Omega = 1$ rad/s and compare the results with your prediction.

(c) Simulate this system for $\sigma_r = 0.1$ and $\Omega = 1$ rad/s and compare the results with your prediction.

Problem 3.21. Consider the LPNI closed loop system of Figure 3.43 with

$$P(s) = \frac{1}{s(s+1)}, \quad C(s) = 2, \tag{3.120}$$

and $F_\Omega(s)$ as given in (3.82).

(a) Compute and plot the random sensitivity function for $f(u) = dz_\Delta(u)$ and $g(y) = y$, where $dz_\Delta(\cdot)$ is defined by (2.23).

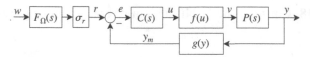

Figure 3.43. Problem 3.21.

(b) Compute and plot the random sensitivity function for $f(u) = u$ and $g(y) =$ $\text{sat}_\alpha(y)$.

3.6 Annotated Bibliography

The material of Section 3.1 is based on the following:

[3.1] Y. Eun, P.T. Kabamba, and S.M. Meerkov, "System types in feedback control with saturating actuators," *IEEE Transactions on Automatic Control*, Vol. 49, pp. 287–291, 2004

The classical notions of system type, overshoot, settling time, bandwidth, resonance peak, and so on, for linear systems, can be found in any undergraduate text on feedback control, for example, see the following:

[3.2] B.C. Kuo, *Automatic Control Systems*, Fifth Edition, Prentice Hall, Englewood Cliffs, NJ, 1987

[3.3] K. Ogata, *Modern Control Engineering*, Second Edition, Prentice Hall, Englewood Cliffs, NJ, 1990

[3.4] R.C. Dorf and R.H. Bishop, *Modern Control Systems*, Eighth Edition, Addison-Wesley, Menlo Park, CA, 1998

[3.5] G.C. Goodwin, S.F. Graebe, and M.E. Salgado, *Control Systems Design*, Prentice Hall, Upper Saddle River, NJ, 2001

[3.6] G.F. Franklin, J.D. Powel, and A. Emami-Naeini, *Feedback Control of Dynamic Systems*, Fourth Edition, Prentice Hall, Englewood Cliffs, NJ, 2002

In this chapter, these notions are generalized for LPNI systems with saturating actuators.
The material of Section 3.2 is based on the following:

[3.7] Y. Eun, P.T. Kabamba, and S.M. Meerkov, "Tracking random references: Random sensitivity function and tracking quality indicators," *IEEE Transactions on Automatic Control*, Vol. 48, pp. 1666–1671, 2003

For Butterworth filters see the following:

[3.8] G. Bianchi and R. Sorrentino, *Electronic Filter Simulation & Design.* McGraw-Hill Professional, 2007

For a description of hard disc servo problem see the following:

[3.9] T.B. Goh, Z. Li, and B.M. Chen, "Design and implementation of a hard disk servo system using robust and perfect tracking approach," *IEEE Transactions on Control Systems Technology*, Vol. 9, 221–233, 2001

The material of Section 3.3 is based on the following:

[3.10] Y. Eun, P.T. Kabamba, and S.M. Meerkov, "Analysis of random reference racking in systems with saturating actuators," *IEEE Transactions on Automatic Control*, Vol. 50, pp. 1861–1866, 2005

The phenomenon of tracking loss, illustrated by the curves (c), (d), and (h) of Figure 3.32 (with infinite roll-off rate), has been discovered experimentally in the following:

[3.11] M. Goldfarb and T. Sirithanapipat, "The effect of actuator saturation on the performance of PD-controlled servo systems," *Mechatronics*, Vol. 9, No. 5, pp. 497–511, 1999

4 Analysis of Disturbance Rejection in LPNI Systems

Motivation: Analysis of disturbance rejection is intended to quantify the standard deviation of the output in closed loop LPNI systems. Unlike the linear case, the usual Lyapunov equation approach cannot be used for this purpose. Therefore, the first goal of this chapter is to provide a quasilinear method applicable to systems with nonlinear instrumentation. While some results in this direction for SISO systems have been described in Chapter 2, we treat here the MIMO case.

Also, we address the problem of fundamental limitations on achievable disturbance rejection in system with saturating actuators. As it is well known, in linear systems with minimum-phase plants, disturbances can be attenuated to any desired level. Clearly, this can not be the case in systems with saturating actuators. Therefore, the second goal of this chapter is to quantify fundamental limitations on disturbance rejection in closed loop LPNI systems and characterize tradeoffs between the authority of the actuators and the achievable disturbance attenuation.

In addition, we present a method for modeling actuators with rate saturation and show that disturbance rejection analysis in LPNI systems with such actuators can be reduced to that with amplitude saturation.

Overview: A quasilinear method for output standard deviation evaluation in closed loop MIMO LPNI systems is developed. The method consists of a Lyapunov equation coupled with transcendental equations defining the quasilinear gains of actuators and sensors. An algorithm for solving these equations with any desired accuracy is provided. Using this method, the achievable level of disturbance rejection as a function of actuator authority is characterized. In addition, a model of actuators with rate saturation is provided, and the disturbance rejection problem with these actuators is discussed.

4.1 Basic Relationships

First, we review the issue of disturbance rejection in SISO LPNI systems and then extend it to the MIMO case.

4.1.1 SISO Systems

The block diagrams of the disturbance rejection LPNI systems considered in this chapter are shown in Figure 4.1. The equations for the quasilinear gains for each of these systems have been derived in Chapter 2. Specifically, they are as follows:

In the case of nonlinear actuator (Figure 4.1(a)):

$$N_a - \mathcal{F}\left(\left\|\frac{F_{\Omega_d}(s)P(s)C(s)}{1+N_a P(s)C(s)}\right\|_2\right) = 0, \tag{4.1}$$

where

$$\mathcal{F}(\sigma) = \int_{-\infty}^{+\infty}\left[\frac{d}{dx}f(x)\right]\frac{1}{\sqrt{2\pi}\sigma}\exp\left(-\frac{x^2}{2\sigma^2}\right)dx. \tag{4.2}$$

In the case of nonlinear sensor (Figure 4.1(b)):

$$N_s - \mathcal{G}\left(\left\|\frac{F_{\Omega_d}(s)P(s)}{1+N_s P(s)C(s)}\right\|_2\right) = 0, \tag{4.3}$$

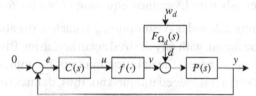

(a) Disturbance rejection with nonlinear actuator

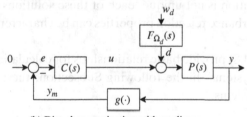

(b) Disturbance rejection with nonlinear sensor

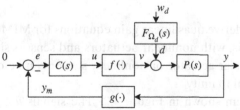

(c) Disturbance rejection with nonlinear actuator and sensor

Figure 4.1. Disturbance rejection LPNI systems.

where

$$G(\sigma) = \int\limits_{-\infty}^{+\infty} \left[\frac{d}{dx} g(x) \right] \frac{1}{\sqrt{2\pi}\sigma} \exp\left(-\frac{x^2}{2\sigma^2}\right) dx. \qquad (4.4)$$

In the case of nonlinear actuator and sensor simultaneously (Figure 4.1(c)):

$$N_a - \mathcal{F}\left(\left\| \frac{F_{\Omega_d}(s)P(s)N_sC(s)}{1 + P(s)N_sC(s)N_a} \right\|_2 \right) = 0, \qquad (4.5)$$

$$N_s - \mathcal{G}\left(\left\| \frac{F_{\Omega_d}(s)P(s)}{1 + P(s)N_sC(s)N_a} \right\|_2 \right) = 0, \qquad (4.6)$$

where $\mathcal{F}$ and $\mathcal{G}$ are defined in (4.2) and (4.4), respectively.

In state space form, the disturbance rejection problem in LPNI systems has been addressed in Subsection 2.2.7 resulting in the quasilinear gain equations

$$N_a = \mathcal{F}\left(\sqrt{C_1 R(N_a, N_s) C_1^T} \right) \qquad (4.7)$$

and

$$N_s = \mathcal{G}\left(\sqrt{C_2 R(N_a, N_s) C_2^T} \right) \qquad (4.8)$$

to be solved simultaneously with Lyapunov equation (2.86) for $R(N_a, N_s)$.

As stated in Theorem 2.2, under Assumption 2.1, each of the above equations has a solution, which can be found, with any desired accuracy, using Bisection Algorithm 2.1. In some special circumstances, the solution may not be unique (see Subsection 2.2.6). However, most often it is indeed unique and, thus, defines the output variance $\sigma_{\hat{y}}$, which can be calculated using a Riccati equation for the resulting quasilinear system. When the solution is not unique, each of these solutions defines an output variance, and the disturbance rejection properties can be characterized by the largest of these variances.

Equations (4.1)–(4.8) provide basic relationships for calculating the output variance in SISO LPNI systems. In the following Subsection, these relationships are extended to MIMO systems.

4.1.2 MIMO Systems

In this subsection, we derive quasilinear gain equations for MIMO LPNI systems in the general case, that is, with nonlinear actuators and sensors simultaneously. The specific cases of nonlinear actuator or nonlinear sensor can be obtained by setting appropriate gains equal to unity.

Consider the system shown in Figure 4.2. The signals $u \in \mathbf{R}^{n_u}$, $y \in \mathbf{R}^{n_y}$, and $y_m \in \mathbf{R}^{n_y}$ are the control input, plant output, and measured output, respectively; $w_1 \in \mathbf{R}^{n_{w1}}$, $w_2 \in \mathbf{R}^{n_{w2}}$, and $w_3 \in \mathbf{R}^{n_{w3}}$ are the disturbance and noise signals, respectively;

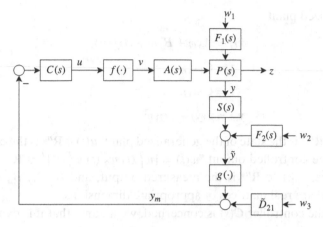

Figure 4.2. Disturbance rejection in MIMO systems with nonlinear actuator and sensor.

$z \in \mathbf{R}^{n_z}$ is the controlled output; $v \in \mathbf{R}^{n_u}$ is the output of the actuator nonlinearity; and $\tilde{y} \in \mathbf{R}^{n_y}$ is the input of the sensor nonlinearity. The transfer matrices $P(s) \in \mathbf{R}^{(n_z+n_y)\times(n_{w_1}+n_u)}$, $C(s) \in \mathbf{R}^{n_u \times n_y}$, $A(s) \in \mathbf{R}^{n_u \times n_u}$, and $S(s) \in \mathbf{R}^{n_y \times n_y}$ represent the plant, controller, actuator, and sensor dynamics, respectively; and $F_1(s) \in \mathbf{R}^{n_{w_1} \times n_{w_1}}$ and $F_2(s) \in \mathbf{R}^{n_y \times n_y}$ represent the coloring filters. The matrix $\tilde{D}_{21}$ represents a direct noise perturbation on the measured output y_m. The functions $f(\cdot)$ and $g(\cdot)$ are the actuator and sensor nonlinearities, respectively.

We assume that w_1, w_2, and w_3 are zero-mean standard uncorrelated Gaussian white noise processes. This does not restrict generality since such normalizations can be carried out by appropriately scaling the coloring filters.

For the nonlinearities $f(u)$ and $g(\tilde{y})$, we assume that

$$f(u) = \begin{bmatrix} f_1(u_1) \\ \vdots \\ f_{n_u}(u_{n_u}) \end{bmatrix} \tag{4.9}$$

and

$$g(\tilde{y}) = \begin{bmatrix} g_1(\tilde{y}_1) \\ \vdots \\ g_{n_y}(\tilde{y}_{n_y}) \end{bmatrix}, \tag{4.10}$$

respectively.

Combining the transfer functions of the plant, actuator, sensor, and coloring filters together, and assuming there are no direct transmission terms from w and u to $\tilde{y}$, z, and y, the system in Figure 4.2, excluding the controller, can be represented by the generalized plant

$$\dot{x}_G = Ax_G + B_1 w + B_2 f(u),$$
$$z = C_1 x_G,$$
$$\tilde{y} = C_2 x_G,$$
$$y_m = g(\tilde{y}) + D_{21} w,$$

(4.11)

where $x_G(t) \in \mathbf{R}^{n_G}$ is the state of the generalized plant, $u(t) \in \mathbf{R}^{n_u}$ is the control input, $z(t) \in \mathbf{R}^{n_z}$ is the controlled output, $w(t) = [w_1^T(t) \, w_2^T(t) \, w_3^T(t)]^T \in \mathbf{R}^{n_{w_1} + n_{w_2} + n_{w_3}}$ is the disturbance, $y_m(t) \in \mathbf{R}^{n_y}$ is the measured output, and A, B_1, B_2, C_1, C_2, and $D_{21} = [0 \, 0 \, \tilde{D}_{21}]$ are real matrices of appropriate dimensions.

As far as the controller $C(s)$ is concerned, we assume that it is dynamic output feedback with realization

$$\dot{x}_C = Mx_C - Ly_m,$$
$$u = Kx_C,$$

(4.12)

where $x_C(t) \in \mathbf{R}^{n_C}$ is the state of the controller and M, L, and K are real matrices of appropriate dimensions.

The disturbance rejection performance of the closed loop system (4.11), (4.12) is measured by

$$\sigma_z = \sqrt{\lim_{t \to \infty} E\{z^T(t)z(t)\}}.$$

(4.13)

Since the steady state covariance matrix of z is given by

$$\Sigma_{zz}^2 = \lim_{t \to \infty} E\{z(t)z^T(t)\},$$

(4.14)

it follows that σ_z can be expressed as

$$\sigma_z = \sqrt{\mathrm{tr}(\Sigma_{zz}^2)},$$

(4.15)

where tr is the trace operator. Introducing the matrix

$$\Sigma_z = \mathrm{diag}\{\Sigma_{zz}\},$$

(4.16)

σ_z can be also expressed as

$$\sigma_z = \sqrt{\mathrm{tr}(\Sigma_z^2)}.$$

(4.17)

With a slight abuse of notations, we will refer to the scalar σ_z as the standard deviation of the vector z.

Combining the generalized plant (4.11) with the output feedback controller (4.12), the closed loop system shown in Figure 4.2 is described by the state space equations

$$\dot{x}_G = Ax_G + B_2 f(Kx_C) + B_1 w,$$
$$\dot{x}_C = Mx_C - Lg(C_2 x_G) - LD_{21}w,$$
$$z = C_1 x_G.$$

(4.18)

Application of stochastic linearization to (4.18) yields the quasilinear system

$$\dot{\hat{x}}_G = A\hat{x}_G + B_2 N_a K\hat{x}_C + B_1 w,$$
$$\dot{\hat{x}}_C = M\hat{x}_C - LN_s C_2 \hat{x}_G - LD_{21} w, \tag{4.19}$$
$$\hat{z} = C_1 \hat{x}_G,$$

where N_a is a diagonal matrix representing the equivalent gains of $f(u)$:

$$N_a = \mathrm{diag}\{\mathcal{F}_1(\sigma_{\hat{u}_1}), \cdots, \mathcal{F}_{n_u}(\sigma_{\hat{u}_{n_u}})\} := \mathcal{F}(\Sigma_{\hat{u}}) \tag{4.20}$$

and N_s is a diagonal matrix representing the equivalent gains of $g(\tilde{y})$:

$$N_s = \mathrm{diag}\{\mathcal{G}_1(\sigma_{\hat{\tilde{y}}_1}), \cdots, \mathcal{G}_{n_y}(\sigma_{\hat{\tilde{y}}_{n_y}})\} := \mathcal{G}(\Sigma_{\hat{\tilde{y}}}). \tag{4.21}$$

Defining

$$\tilde{A} = \begin{bmatrix} A & 0 \\ 0 & M \end{bmatrix}, \quad \tilde{N} = \begin{bmatrix} N_a & 0 \\ 0 & N_s \end{bmatrix},$$
$$\tilde{B}_1 = \begin{bmatrix} B_1 \\ -LD_{21} \end{bmatrix}, \quad \tilde{C}_1 = \begin{bmatrix} C_1 & 0 \end{bmatrix}, \tag{4.22}$$
$$\tilde{B}_2 = \begin{bmatrix} B_2 & 0 \\ 0 & -L \end{bmatrix}, \quad \tilde{C}_2 = \begin{bmatrix} 0 & K \\ C_2 & 0 \end{bmatrix},$$

the system in (4.19) can be finally written as

$$\dot{\hat{x}} = (\tilde{A} + \tilde{B}_2 \tilde{N} \tilde{C}_2)\hat{x} + \tilde{B}_1 w,$$
$$\hat{z} = \tilde{C}_1 \hat{x}, \tag{4.23}$$

where $\hat{x} = \begin{bmatrix} \hat{x}_G^T & \hat{x}_C^T \end{bmatrix}^T$ is the state of the plant and controller.

Theorem 4.1. *Consider the LPNI system (4.18) and assume that the processes u, y, and z are stationary. Then, the stochastic linearization estimate of the standard deviation of z is*

$$\sigma_{\hat{z}} = \sqrt{\mathrm{tr}(\tilde{C}_1 \tilde{P} \tilde{C}_1^T)}, \tag{4.24}$$

where $\tilde{P}$ together with $\tilde{N}$ satisfy the system of equations

$$(\tilde{A} + \tilde{B}_2 \tilde{N} \tilde{C}_2)\tilde{P} + \tilde{P}(\tilde{A} + \tilde{B}_2 \tilde{N} \tilde{C}_2)^T + \tilde{B}_1 \tilde{B}_1^T = 0, \tag{4.25}$$

$$\tilde{N} - \mathrm{diag}\{\mathcal{F}(\Sigma_{\hat{u}}), \mathcal{G}(\Sigma_{\hat{\tilde{y}}})\} = 0, \tag{4.26}$$

provided that these equations have a unique solution.

Proof. See Section 8.3.

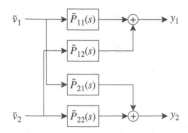

Figure 4.3. MIMO plant for Example 4.1.

A multivariable bisection algorithm analogous to Bisection Algorithm 2.1 can be used to solve the above system of equations. An alternative algorithm to accomplish this is the multivariable Newton iteration.

Example 4.1. Consider the two-input two-output plant shown in Figure 4.3, where $\tilde{v}_1$ and $\tilde{v}_2$ are the inputs and y_1 and y_2 are the outputs, and assume that

$$\tilde{P}_{11}(s) = \frac{1}{s+1}, \ \tilde{P}_{12}(s) = \frac{0.1}{s+1}, \ \tilde{P}_{21}(s) = \frac{0.15}{2.5s+1}, \ \tilde{P}_{22}(s) = \frac{1.5}{2.5s+1}. \tag{4.27}$$

Thus,

$$\tilde{P}(s) = \begin{bmatrix} \tilde{P}_{11}(s) & \tilde{P}_{12}(s) \\ \tilde{P}_{21}(s) & \tilde{P}_{22}(s) \end{bmatrix} = \begin{bmatrix} \dfrac{1}{s+1} & \dfrac{0.1}{s+1} \\ \dfrac{0.15}{2.5s+1} & \dfrac{1.5}{2.5s+1} \end{bmatrix}. \tag{4.28}$$

A Gilbert realization of $\tilde{P}(s)$ is

$$\begin{bmatrix} \dot{x}_{\tilde{P}_1} \\ \dot{x}_{\tilde{P}_2} \end{bmatrix} = \begin{bmatrix} -1 & 0 \\ 0 & -0.4 \end{bmatrix} \begin{bmatrix} x_{\tilde{P}_1} \\ x_{\tilde{P}_2} \end{bmatrix} + \begin{bmatrix} 1 & 0 \\ 0 & 1 \end{bmatrix} \begin{bmatrix} \tilde{v}_1 \\ \tilde{v}_2 \end{bmatrix},$$
$$\begin{bmatrix} y_1 \\ y_2 \end{bmatrix} = \begin{bmatrix} 1 & 0.1 \\ 0.06 & 0.6 \end{bmatrix} \begin{bmatrix} x_{\tilde{P}_1} \\ x_{\tilde{P}_2} \end{bmatrix}, \tag{4.29}$$

where $x_{\tilde{P}_1}$ and $x_{\tilde{P}_2}$ are the states of the plant.

Assume that the plant is controlled by two saturating actuators, and the plant inputs are disturbed by input disturbances as shown in Figure 4.4. The signals u_1 and u_2 are the inputs of the actuators; v_1 and v_2 are the outputs of the actuators; and d_1 and d_2 are the disturbances. The disturbances d_1 and d_2 are modeled as colored processes obtained by filtering the standard white noise processes w_{11} and w_{12} through the filters $F_{11}(s)$ and $F_{12}(s)$, respectively. For simplicity, it is assumed that

$$F_{11}(s) = \frac{5/2}{5s+1} \tag{4.30}$$

and

$$F_{12}(s) = \frac{5/3}{10/3s+1}. \tag{4.31}$$

Moreover, the actuator dynamics are assumed fast enough so that they can be neglected (i.e., $A_1(s) = 1$ and $A_2(s) = 1$). Note that a realization of $F_{11}(s)$ is

$$\dot{x}_{F_1} = -0.2x_{F_1} + w_{11},$$
$$d_1 = 0.5x_{F_1},$$
(4.32)

where x_{F_1} is the state of the filter $F_{11}(s)$, and a realization of $F_{12}(s)$ is

$$\dot{x}_{F_2} = -0.3x_{F_2} + w_{12},$$
$$d_2 = 0.5x_{F_2},$$
(4.33)

where x_{F_2} is the state of the filter $F_{12}(s)$.

Next, assume that y_1 and y_2 are measured using two saturating sensors, and the sensor outputs are corrupted by noise processes n_1 and n_2, resulting in measured outputs y_{m_1} and y_{m_2}, as shown in Figure 4.4. For simplicity, we assume that

$$n_1 = 0.04w_{31}$$
(4.34)

and

$$n_2 = 0.03w_{32}.$$
(4.35)

Finally, to quantify disturbance rejection performance of the system, the controlled outputs z_1 and z_2 are introduced as shown in Figure 4.4. Thus,

$$z_1 = y_1$$
(4.36)

and

$$z_2 = y_2.$$
(4.37)

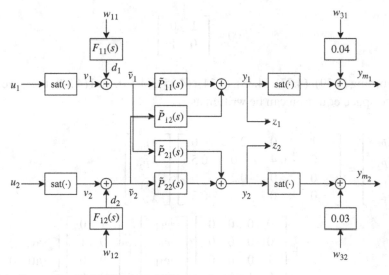

Figure 4.4. Generalized MIMO plant for Example 4.1.

Note that the standard deviation of $z = \begin{bmatrix} z_1 & z_2 \end{bmatrix}^T$ is given by

$$\sigma_z = \sqrt{\sigma_{z_1}^2 + \sigma_{z_2}^2}. \tag{4.38}$$

The transfer matrices in Figure 4.2 can be easily calculated from the data presented above. Indeed, these transfer matrices are

$$P(s) = \begin{bmatrix} \tilde{P}_{11}(s) & \tilde{P}_{12}(s) & \tilde{P}_{11}(s) & \tilde{P}_{12}(s) \\ \tilde{P}_{21}(s) & \tilde{P}_{22}(s) & \tilde{P}_{21}(s) & \tilde{P}_{22}(s) \\ \tilde{P}_{11}(s) & \tilde{P}_{12}(s) & \tilde{P}_{11}(s) & \tilde{P}_{12}(s) \\ \tilde{P}_{21}(s) & \tilde{P}_{22}(s) & \tilde{P}_{21}(s) & \tilde{P}_{22}(s) \end{bmatrix}, \tag{4.39}$$

$$F_1(s) = \begin{bmatrix} F_{11}(s) & 0 \\ 0 & F_{12}(s) \end{bmatrix}, \tag{4.40}$$

$$F_2(s) = \begin{bmatrix} F_{21}(s) & 0 \\ 0 & F_{22}(s) \end{bmatrix}, \tag{4.41}$$

$$A(s) = \begin{bmatrix} 1 & 0 \\ 0 & 1 \end{bmatrix}, \tag{4.42}$$

and

$$S(s) = \begin{bmatrix} 1 & 0 \\ 0 & 1 \end{bmatrix}. \tag{4.43}$$

Combining (4.29), (4.32), (4.33), (4.34), (4.35), (4.36), and (4.37), the generalized plant state space equation can be written as

$$\begin{bmatrix} \dot{x}_{\tilde{P}_1} \\ \dot{x}_{\tilde{P}_2} \\ \dot{x}_{\tilde{F}_1} \\ \dot{x}_{\tilde{F}_2} \end{bmatrix} = \begin{bmatrix} -1 & 0 & 0.5 & 0 \\ 0 & -0.4 & 0 & 0.5 \\ 0 & 0 & -0.2 & 0 \\ 0 & 0 & 0 & -0.3 \end{bmatrix} \begin{bmatrix} x_{\tilde{P}_1} \\ x_{\tilde{P}_2} \\ x_{\tilde{F}_1} \\ x_{\tilde{F}_2} \end{bmatrix}$$
$$+ \begin{bmatrix} 0 & 0 & 0 & 0 \\ 0 & 0 & 0 & 0 \\ 1 & 0 & 0 & 0 \\ 0 & 1 & 0 & 0 \end{bmatrix} \begin{bmatrix} w_{11} \\ w_{12} \\ w_{21} \\ w_{22} \end{bmatrix} + \begin{bmatrix} 1 & 0 \\ 0 & 1 \\ 0 & 0 \\ 0 & 0 \end{bmatrix} \begin{bmatrix} \text{sat}(u_1) \\ \text{sat}(u_2) \end{bmatrix},$$

$$\begin{bmatrix} z_1 \\ z_2 \end{bmatrix} = \begin{bmatrix} 1 & 0.1 & 0 & 0 \\ 0.06 & 0.6 & 0 & 0 \end{bmatrix} \begin{bmatrix} x_{\tilde{P}_1} \\ x_{\tilde{P}_2} \\ x_{\tilde{F}_1} \\ x_{\tilde{F}_2} \end{bmatrix},$$

$$\begin{bmatrix} y_{m1} \\ y_{m2} \end{bmatrix} = \begin{bmatrix} \mathrm{sat}(y_1) \\ \mathrm{sat}(y_2) \end{bmatrix} + \begin{bmatrix} 0 & 0 & 0.04 & 0 \\ 0 & 0 & 0 & 0.03 \end{bmatrix} \begin{bmatrix} w_{11} \\ w_{12} \\ w_{21} \\ w_{22} \end{bmatrix},$$

$$\begin{bmatrix} y_1 \\ y_2 \end{bmatrix} = \begin{bmatrix} 1 & 0.1 & 0 & 0 \\ 0.06 & 0.6 & 0 & 0 \end{bmatrix} \begin{bmatrix} x_{\tilde{P}_1} \\ x_{\tilde{P}_2} \\ x_{\tilde{F}_1} \\ x_{\tilde{F}_2} \end{bmatrix}, \tag{4.44}$$

which is in the form (4.11).

The open loop standard deviation of z is $\sigma_z = 1.0360$. By ignoring the saturations both in the actuators and sensors, a controller is designed to attenuate σ_z to approximately 10% of the open loop value. The resulting controller has the form (4.12) with

$$M = \begin{bmatrix} -103.7154 & -8.6974 & 0.0061 & -0.0005 \\ -8.7844 & -64.0333 & -0.0009 & 0.0059 \\ -22.9993 & -2.0007 & -0.2000 & 0.0000 \\ -2.3499 & -17.5289 & 0.0000 & -0.3000 \end{bmatrix},$$

$$L = \begin{bmatrix} -3.9106 & 0.3178 \\ 0.0936 & -6.3501 \\ -23.0296 & 0.5038 \\ -0.6031 & -29.1144 \end{bmatrix}, \tag{4.45}$$

$$K = \begin{bmatrix} -98.8238 & -8.4970 & -0.4939 & -0.0005 \\ -8.4970 & -59.8326 & -0.0009 & -0.4941 \end{bmatrix},$$

and achieves $\sigma_z = 0.0994$ in the absence of saturations. To investigate the performance degradation in the presence of saturating actuators and sensors, we apply the multivariable Newtown algorithm, which yields $\sigma_{\hat{z}} = 0.1638$, that is, 65% degradation. Using simulation, the actual value of the standard deviation of z is calculated as $\sigma_z = 0.167$. Clearly, the predicted standard deviation agrees well with its actual value. The equivalent gains are

$$N_a = \begin{bmatrix} 0.2495 & 0 \\ 0 & 0.3957 \end{bmatrix}, \quad N_s = \begin{bmatrix} 1.0000 & 0 \\ 0 & 1.0000 \end{bmatrix} \tag{4.46}$$

and they indicate that the actuators saturate quite often, while the sensors do not. As one can see, the linear design does not work well with nonlinear instrumentation, and to ensure the desired performance, either the controller or the instrumentation must be modified. These issues are addressed in Chapters 6 and 7.

4.2 Fundamental Limitations on Disturbance Rejection

The disturbance rejection capabilities of linear feedback systems are well known: If the plant is minimum phase, the output variance can be reduced to any desired level; if the plant is non-minimum phase, the disturbance rejection capabilities are limited, that is, there exists a fundamental bound on the achievable disturbance rejection defined by the locations of the plant non-minimum phase zeros.

What are disturbance rejection capabilities of LPNI systems? This is the question addressed in this section. Intuitively, it is clear that these capabilities may be limited even for minimum phase plants. Indeed, if the actuator is, for example, saturating, bounded inputs might not be sufficient to combat disturbances, leading to an effect similar to non-minimum phase plants. Below, we provide a quantification of this phenomenon. The approach is based on linear matrix inequalities (LMI).

Consider the SISO LPNI system of Figure 4.5. It differs from the standard configuration of Figure 4.1(a) in that the full state vector, x, is used for feedback (rather than the scalar output, y). If one proves that there is a finite bound on the disturbance rejection capabilities under state space feedback, Kx, where K is the vector of feedback gains, one would conjecture that this bound holds for any output feedback, including dynamic ones. For simplicity, we assume also that the disturbance is the (unfiltered) white noise process.

The state space description of this system can be given as follows:

$$\dot{x} = Ax + B\text{sat}_\alpha(Kx) + Bw,$$
$$y = Cx, \tag{4.47}$$

where $x \in \mathbf{R}^n$ is the state and constant matrices A, B, and C are of appropriate dimensions. The stochastically linearized version of this system is

$$\dot{\hat{x}} = (A + BNK)\hat{x} + Bw,$$
$$\hat{y} = C\hat{x}, \tag{4.48}$$

where

$$N - \text{erf}\left(\frac{\alpha}{\sqrt{2}\sigma_{\hat{u}}}\right), \tag{4.49}$$

$$\sigma_{\hat{u}} = \sqrt{KPK^T}, \tag{4.50}$$

and $P = P^T > 0$ satisfies

$$(A + BNK)P + P(A + BNK)^T + BB^T = 0. \tag{4.51}$$

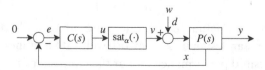

Figure 4.5. LPNI disturbance rejection systems with state space feedback.

While the output variance of this system, $\sigma_{\hat{y}}$, can be analyzed using the method of Section 2.2, it can be also investigated using the LMI approach, as stated in the following:

Theorem 4.2. *System (4.48) has $\sigma_{\hat{y}}$ less than or equal to γ if and only if there exists $Q \in \mathbf{R}^{n \times n}$, $Q = Q^T > 0$ such that*

$$(A + BNK)Q + Q(A + BNK)^T + BB^T \leq 0,$$
$$CQC^T - \gamma^2 \leq 0. \tag{4.52}$$

Proof. See Section 8.3.

In addition, the LMI approach can be used to investigate fundamental limitations of the disturbance rejection. This is based on the following:

Theorem 4.3. *Consider system (4.48) subject to (4.49)–(4.51). There exists K such that $\sigma_{\hat{y}}$ is less than or equal to γ if and only if there exists $Y \in \mathbf{R}^{1 \times n}$ and $\bar{P} \in \mathbf{R}^{n \times n}$, $\bar{P} = \bar{P}^T > 0$ such that*

$$\begin{bmatrix} \dfrac{2\alpha^2}{\pi} & Y \\ Y^T & \bar{P} \end{bmatrix} > 0, \tag{4.53}$$

$$A\bar{P} + \bar{P}A^T + BY + Y^T B^T + BB^T \leq 0, \tag{4.54}$$

$$C\bar{P}C^T - \gamma^2 \leq 0. \tag{4.55}$$

Proof. See Section 8.3.

Using this theorem, the saturation level α can be minimized subject to the LMI (4.53)–(4.55) for a given performance specification γ. This allows us to construct a tradeoff locus, or feasibility boundary in the (γ, α)-plane, below which it is impossible to find a state feedback that satisfies the given level of performance for a given saturation level.

To illustrate the tradeoff locus, consider system (4.48) with

$$A = \begin{bmatrix} 0 & 1 \\ -2 & -3 \end{bmatrix}, \quad B = \begin{bmatrix} 0 \\ 1 \end{bmatrix}, \quad C = [1 - 2]. \tag{4.56}$$

Then using (4.53)–(4.55) and the MATLAB LMI toolbox, we calculate the minimum achievable α, α_{min}, for $\gamma \in [0.1, 1]$. The result is shown in Figure 4.6. As expected, α_{min} is a monotonically decreasing function of γ, with $\alpha_{min} = 0$ for $\gamma \geq \gamma_{OL}$, where γ_{OL} is the open loop $\sigma_{\hat{y}}$.

4.3 LPNI Systems with Rate-Saturated Actuators

In this section, we model rate-saturated actuators, define the notion of their bandwidth, and develop a method for LPNI system analysis with such actuators. The latter

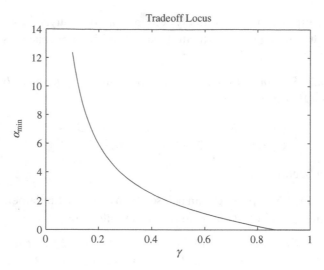

Figure 4.6. Tradeoff locus for example system (4.56).

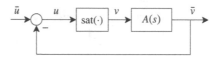

Figure 4.7. Rate-saturated actuator model.

is accomplished by reducing the closed loop LPNI system with rate-saturated actuator to an equivalent LPNI system with amplitude saturation.

4.3.1 Modeling Rate-Saturated Actuators

Sometimes, actuators are used in a servo configuration to obtain a desired dynamic characteristic, and to reduce the nonlinear effects due to friction and material properties. Assuming that the sensor used in forming the servo connection has unity gain, such a servo actuator can be modeled as the block diagram shown in Figure 4.7, where $\bar{u}$ and $\bar{v}$ are the input and output of the servo, respectively. In the control literature, the actuator model shown in Figure 4.7 is often referred to as a rate-saturated or rate-limited actuator, especially when $A(s) = 1/s$. To simplify the presentation, we assume throughout this section that

$$A(s) = \frac{1}{s}.\tag{4.57}$$

To illustrate the effect of rate saturation, the system in Figure 4.7 is simulated using a triangular input with three frequencies 0.01 Hz, 0.1 Hz, and 1 Hz. The output of the system, together with its input, are shown in Figure 4.8. Obviously, as the frequency of the input increases, the effect of rate saturation becomes more pronounced.

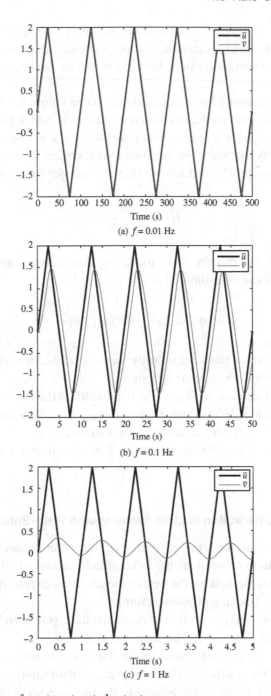

Figure 4.8. Response of a rate-saturated actuator.

4.3.2 Bandwidth of Rate-Saturated Actuators

In the control literature, it is common to refer to the bandwidth of actuators that
are subject to saturation. However, what is meant by this is almost never specified.

Below, we use the method of stochastic linearization to give a bandwidth definition for systems subject to saturation. Although the definition is applicable to any given system, the presentation can be best illustrated by the rate-saturated actuator described above.

Consider the rate-limited actuator model shown in Figure 4.7. The bandwidth of this system can be defined as the maximum frequency at which the output $\bar{v}$ agrees with the input $\bar{u}$ in a certain sense. Assuming that $\bar{u}$ is a zero-mean wss Gaussian process, we can apply stochastic linearization to this system. Denoting the stochastic linearization estimate of $\bar{v}$ by $\hat{\bar{v}}$, it follows that the transfer function from $\bar{u}$ to $\hat{\bar{v}}$ is

$$H(s) = \frac{N}{s+N}, \tag{4.58}$$

where N is the equivalent gain of $\text{sat}(u)$. Using this transfer function, the power spectral density of $\bar{v}$ can be estimated as

$$S_{\bar{v}}(\omega) \simeq S_{\hat{\bar{v}}}(\omega) = |H(j\omega)|^2 S_{\bar{u}}(\omega). \tag{4.59}$$

Hence, the bandwidth of the stochastically linearized system is an estimate of the bandwidth of the original nonlinear system.

Based on this discussion, we define the bandwidth of the rate-saturated actuator as the 3dB bandwidth of $H(s)$, which is equal to $\omega_{BN} = N$. Note that the introduced bandwidth ω_{BN} is a decreasing function of the variance of $\bar{u}$.

The bandwidth of any linear system with nonlinear instrumentation can be defined similarly.

4.3.3 Disturbance Rejection in LPNI Systems with Rate-Saturated Actuators

An LPNI system containing a rate-saturated actuator can be easily converted into an equivalent system that contains only an amplitude-saturated actuator. Below, we illustrate this analysis method for the case of the disturbance rejection problem. The development for the tracking problem is similar.

Consider the plant $P(s)$ controlled by the controller $C(s)$ through a rate-saturated actuator as shown in Figure 4.9. The signals $w \in \mathbf{R}^{n_w}$, $\bar{u} \in \mathbf{R}$, $u \in \mathbf{R}$, $v \in \mathbf{R}$, $\bar{v} \in \mathbf{R}$, $z \in \mathbf{R}^{n_z}$, and $y \in \mathbf{R}$ are the disturbance/noise, input to the rate-saturated actuator, input of the saturation, output of the saturation, control input, controlled output, and measured output, respectively. Similar to the previous sections, we assume that the plant $P(s)$ has the representation

$$\begin{aligned}
\dot{x}_G &= Ax_G + B_1 w + B_2 \bar{v}, \\
z &= C_1 x_G + D_{12} \bar{u}, \\
y &= C_2 x_G + D_{21} w,
\end{aligned} \tag{4.60}$$

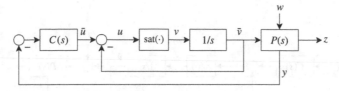

Figure 4.9. Disturbance rejection with rate-saturated actuator.

where $x_G \in \mathbf{R}^{n_{x_G}}$ is the state of the plant, and the controller $C(s)$ has the representation

$$\dot{x}_C = Mx_C - Ly,$$
$$\bar{u} = Kx_C,$$

(4.61)

where $x_C \in \mathbf{R}^{n_{x_C}}$ is the state of the controller.

Combining the above equations, it follows that

$$\dot{x}_G = Ax_G + B_2 x_A + B_1 w,$$
$$\dot{x}_C = Mx_C - LC_2 x_G - LD_{21} w,$$
$$\dot{x}_A = f(Kx_C - x_A),$$
$$z = C_1 x_G + D_{12} Kx_C,$$
$$u = Kx_C - x_A,$$

(4.62)

where $x_A \in \mathbf{R}$ is the state of the rate-saturated actuator. Applying the method of stochastic linearization, we obtain a set of equations similar to (4.23):

$$\begin{bmatrix} \dot{\hat{x}}_G \\ \dot{\hat{x}}_C \\ \dot{\hat{x}}_A \end{bmatrix} = \begin{bmatrix} A & 0 & B_2 \\ -LC_2 & M & 0 \\ 0 & NK & -N \end{bmatrix} \begin{bmatrix} \hat{x}_G \\ \hat{x}_C \\ \hat{x}_A \end{bmatrix} + \begin{bmatrix} B_1 \\ -LD_{21} \\ 0 \end{bmatrix} w,$$

$$\hat{z} = \begin{bmatrix} C_1 & D_{12}K & 0 \end{bmatrix} \begin{bmatrix} \hat{x}_G \\ \hat{x}_C \\ \hat{x}_A \end{bmatrix},$$

(4.63)

$$\hat{u} = \begin{bmatrix} 0 & K & -1 \end{bmatrix} \begin{bmatrix} \hat{x}_G \\ \hat{x}_C \\ \hat{x}_A \end{bmatrix},$$

where N can be computed as in Theorem 4.1. Hence, systems with rate-saturated actuators can be analyzed using the method developed above for systems with amplitude-saturated actuators.

Example 4.2. Consider the control system shown in Figure 4.10 and assume that

$$P(s) = \frac{1}{s+1}, \quad F(s) = \frac{2}{s+1}, \quad C(s) = \frac{1.5}{0.2s+1}.$$

(4.64)

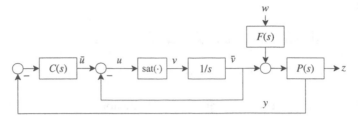

Figure 4.10. System for Example 4.2.

Combining the states of all subsystems and letting $z = y$, this system can be put into the standard state space form as

$$
\dot{x}_G = \begin{bmatrix} -1 & 0 & 1 & 1 \\ -7.5 & -5 & 0 & 0 \\ 0 & 0 & 0 & 0 \\ 0 & 0 & 0 & -1 \end{bmatrix} x_G + \begin{bmatrix} 0 \\ 0 \\ 0 \\ 2 \end{bmatrix} w + \begin{bmatrix} 0 \\ 0 \\ 1 \\ 0 \end{bmatrix} \text{sat}(u),
$$

$$
z = \begin{bmatrix} 1 & 0 & 0 & 0 \end{bmatrix} x_G,
$$

$$
y = \begin{bmatrix} 1 & 0 & 0 & 0 \end{bmatrix} x_G,
$$

$$
u = \begin{bmatrix} 0 & 1 & -1 & 0 \end{bmatrix} x_G.
$$

(4.65)

The open loop standard deviation of z is 1. Applying Bisection Algorithm 2.1 with $\epsilon = 1 \times 10^{-6}$ to this system, we obtain $\sigma_{\hat{z}} = 0.7071$ and $N = 0.8247$. Using simulation, standard deviation of z is estimated as $\sigma_z = 0.719$. For comparison, if the actuator is used as an amplitude-saturated actuator instead of rate-saturated one, then it turns out that $\sigma_{\hat{z}} = 0.9406$ and $N = 0.5326$. The time traces of y and u for both cases are shown in Figure 4.11.

Clearly, in this example, the performance with the rate-saturated actuator is better than that with the amplitude-saturated actuator. However, this is not always the case. To illustrate this point, the filter $F(s)$ is changed to

$$
F(s) = \frac{4.69}{10s + 1}. \tag{4.66}
$$

Again, the open loop standard deviation of z is 1, but $\sigma_{\hat{z}} = 0.4450$ with the rate-saturated actuator and $\sigma_{\hat{z}} = 0.3146$ with the amplitude-saturated actuator. The corresponding N values are $N = 0.9999$ and $N = 0.9702$, respectively. The time traces of y and u are shown in Figure 4.12.

4.4 Summary

- The disturbance rejection problem in MIMO LPNI systems can be solved using a Ricatti equation coupled with transcendental equations for the quasilinear gains.

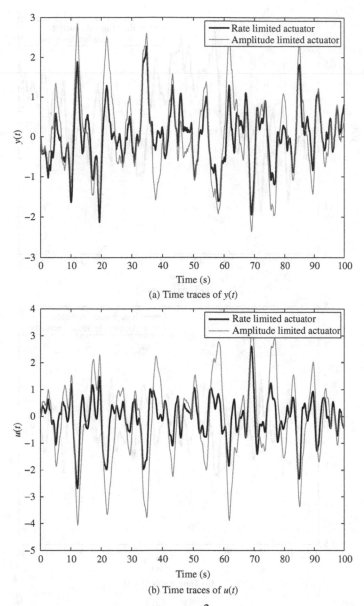

Figure 4.11. Time traces of $y(t)$ and $u(t)$, $F(s) = \dfrac{2}{s+1}$.

- The disturbance rejection capabilities of LPNI systems with saturating actuators are fundamentally limited, even when the plant is minimum phase. These limitations can be quantified using a linear matrix inequality approach.
- The analysis of LPNI systems with rate saturation can be reduced to the analysis of LPNI systems with amplitude saturation.
- The material included in this chapter provides a solution of the Analysis problem introduced in Section 1.2 as far as disturbance rejection is concerned.

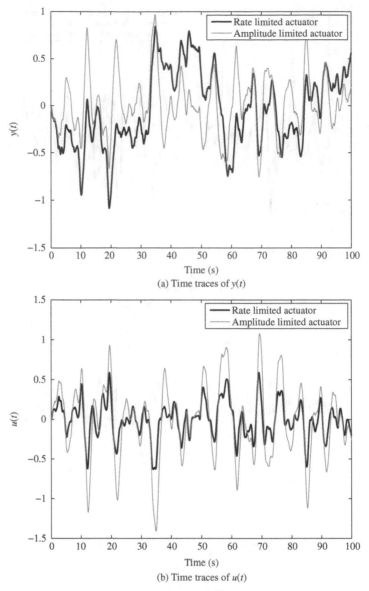

Figure 4.12. Time traces of $y(t)$ and $u(t)$, $F(s) = \dfrac{4.69}{10s+1}$.

4.5 Problems

Problem 4.1. Consider the system of Example 4.1 with the only difference that the plant is given by

$$\tilde{P}_{11}(s) = \frac{1}{2s+1}, \ \tilde{P}_{12}(s) = \frac{0.1}{2s+1}, \ \tilde{P}_{21}(s) = \frac{0.15}{5s+1}, \ \tilde{P}_{22}(s) = \frac{1.5}{5s+1}.$$

(a) Calculate the open loop standard deviation of z

(b) Assuming that the actuators and the sensors are linear, design a linear controller that attenuates σ_z to about 10% on the open loop value.

(c) Assuming that the actuators and the sensors are saturating as in Figure 4.4, calculate $\sigma_{\hat{z}}$ and comment on the resulting performance degradation.

Problem 4.2. Consider the system (4.47) with A, B, and C given by:

$$A = \begin{bmatrix} 0 & 1 \\ -5 & -2 \end{bmatrix}, \quad B = \begin{bmatrix} 0 \\ 1 \end{bmatrix}, \quad C = 25.$$

(a) Using Theorem 4.3, analyze the fundamental limitations on disturbance rejection when the actuator is the standard saturating function.

(b) Calculate the trade-off locus for this system (similar to that of Figure 4.6).

Problem 4.3. Consider the system of Example 2.3 and assume that the rate-saturated actuator of Figure 4.7 is used instead of the amplitude-saturated actuator of Figure 2.8(a). Based on the method of Section 4.3, calculate the standard deviation $\sigma_{\hat{y}}$ and compare it with that derived in Example 2.3.

4.6 Annotated Bibliography

The material of Section 4.1 is based on the following:

[4.1] C. Gokcek, P.T. Kabamba and S.M. Meerkov, "Disturbance rejection in control systems with saturating actuators," *Nonlinear Analysis – Theory, Methods & Applications*, Vol. 40, pp. 213–226, 2000

For the analysis of fundamental limitations on disturbance rejection in LPNI systems using LMIs, see the following:

[4.2] P.T. Kabamba, S.M. Meerkov and C.N. Takahashi, "LMI approach to disturbance rejection in systems with saturating actuators," *University of Michigan Control Group Report*, No. CGR10-03, 2010

5 Design of Reference Tracking Controllers for LPNI Systems

Motivation: In designing linear feedback systems, tracking quality specifications are typically mapped into admissible domains in the complex plane and then the root locus technique is used to synthesize a controller that places closed loop poles in the desired locations. This methodology motivates the first goal of this chapter: to generalize the notion of admissible domains and the root locus technique to LPNI systems. Accomplishing this goal leads to a quasilinear method for PID controller design.

Overview: A technique for mapping the random reference tracking quality indicators, introduced in Chapter 3, into admissible domains on the complex plane is introduced (Section 5.1). Next, the root locus method for systems with saturating actuators is developed (S-root locus, Section 5.2). It turns out that the S-root locus is a subset of the usual root locus typically terminating prior to the open loop zeros. A method for calculating these termination points is provided. In addition, we equip the S-root locus with the so-called truncation points, which define its segments where tracking without amplitude truncation is possible.

5.1 Admissible Pole Locations for Random Reference Tracking

5.1.1 Scenario

In linear Control Theory, admissible domains for tracking deterministic references are defined using the classical "quality indicators," that is, overshoot and settling, rise, and delay times. In this section, we develop a similar approach for tracking random references. The system addressed is shown in Figure 5.1, where, as before, $P(s)$ and $C(s)$ are the plant and controller, respectively, and $F_\Omega(s)$ is the coloring filter.

More specifically, instead of the overshoot, settling time, and so on, we use the quality indicators for tracking random references introduced in Section 3.2:

$$I_1 = R_{dc} \tag{5.1}$$

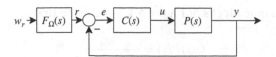

Figure 5.1. Tracking control system with a random reference.

$$I_2 = \frac{\Omega}{R\Omega_{BW}}, \tag{5.2}$$

$$I_3 = \min\left\{RM_r - 1, \frac{\Omega}{R\Omega_r}\right\}. \tag{5.3}$$

Here R_{dc}, $R\Omega_{BW}$, RM_r, and $R\Omega_r$ are the d.c. gain, bandwidth, resonance peak and resonance frequency of the random sensitivity function, that is,

$$R_{dc} = \lim_{\Omega \to 0} RS(\Omega), \tag{5.4}$$

$$R\Omega_{BW} = \min\{\Omega : RS(\Omega) = 1/\sqrt{2}\}, \tag{5.5}$$

$$R\Omega_r = \arg\sup_{\Omega>0} RS(\Omega), \tag{5.6}$$

$$RM_r = \sup_{\Omega>0} RS(\Omega), \tag{5.7}$$

where

$$RS(\Omega) = \|F_\Omega(s)S(s)\|_2 \tag{5.8}$$

if $\|F_\Omega(s)\|_2 = 1$ or

$$RS(\Omega) = \frac{\|F_\Omega(s)S(s)\|_2}{\|F_\Omega(s)\|_2} \tag{5.9}$$

if $\|F_\Omega(s)\|_2 \neq 1$, and $S(s)$ is the usual sensitivity function:

$$S(s) = \frac{1}{1 + C(s)P(s)}. \tag{5.10}$$

As discussed in Chapter 3, I_1 describes static unresponsiveness; I_2 characterizes dynamics of the response (large I_2 implies lagging or oscillatory behavior); and I_3 discriminates between the latter two phenomena (large I_3 implies excessive oscillations). Indicators I_1–I_3 can be evaluated either numerically or analytically as described in Subsection 3.2.4.

As for the dynamic blocks of Figure 5.1, we assume that

$$C(s)P(s) = \frac{\omega_n^2}{s(s + 2\zeta\omega_n)} \tag{5.11}$$

and

$$F_\Omega(s) = \sqrt{\frac{3}{\Omega}} \left(\frac{\Omega^3}{s^3 + 2\Omega s^2 + 2\Omega^2 s + \Omega^3} \right). \tag{5.12}$$

Due to (5.11), the closed loop transfer function in Figure 5.1 is

$$T(s) = \frac{\omega_n^2}{s^2 + 2\zeta\omega_n s + \omega_n^2}, \tag{5.13}$$

that is, a prototype second order system with natural frequency ω_n and damping ratio ζ. Since the sensitivity function for this system is

$$S(s) = \frac{s^2 + 2\zeta\omega_n s}{s^2 + 2\zeta\omega_n s + \omega_n^2}, \tag{5.14}$$

it follows from (5.9) that the random sensitivity function is a function of ω_n and ζ as well as Ω, that is,

$$RS = RS(\Omega, \omega_n, \zeta). \tag{5.15}$$

Moreover, it is easy to see that RS depends, in fact, not on ω_n and Ω separately but on their ratio, that is,

$$RS(\Omega, \omega_n, \zeta) = RS\left(\frac{\Omega}{\omega_n}, \zeta\right). \tag{5.16}$$

Indeed, using the substitution

$$\omega = \Omega\hat{\omega}, \tag{5.17}$$

expression (5.8) can be rewritten as

$$RS(\Omega/\omega_n) = \sqrt{\int_{-\infty}^{\infty} \frac{3}{2\pi} \left| \frac{\Psi(j\hat{\omega})}{(1 - 2\hat{\omega}^2) + j(\hat{\omega}^3 + 2\hat{\omega})} \right|^2 d\hat{\omega}}, \tag{5.18}$$

where

$$\Psi(j\hat{\omega}) = \frac{-\rho^2\hat{\omega}^2 + j2\zeta\rho\hat{\omega}}{(1 - \rho^2\hat{\omega}^2) + j2\zeta\rho\hat{\omega}} \tag{5.19}$$

and

$$\rho = \frac{\Omega}{\omega_n}. \tag{5.20}$$

The quantity Ω/ω_n is referred to as the *dimensionless bandwidth*. Below, we denote the random sensitivity function either as $RS(\Omega)$ or as $RS(\Omega/\omega_n)$, depending on the issue at hand.

5.1.2 Admissible Domains for Random Reference Tracking by Prototype Second Order System

As it follows from the above, the admissible pole domain for tracking random references is the intersection of two sets in the s-plane, defined by the inequalities

$$I_2 \leq \gamma, \tag{5.21}$$

$$I_3 \leq \eta, \tag{5.22}$$

where γ and η are sufficiently small positive constants. Clearly, the boundaries of these sets are level curves of I_2 and I_3. Below, these level curves are constructed.

Admissible domains from the point of view of I_2: Assume that the closed loop transfer function (5.13) has poles $s_1, s_2 = \sigma \pm j\omega$, $\sigma < 0$. We are interested in studying the behavior of I_2 as a function of σ and ω.

In order to make the level curves of I_2 independent of Ω, using the normalization introduced in Subsection 5.1.1, we view I_2 as a function of "dimensionless" pole locations $(\sigma/\Omega) \pm j(\omega/\Omega)$. Figure 5.2 depicts these level curves, calculated using the method described in Subsection 3.2.3.

Thus, all poles located to the left of the curve $I_2 = \gamma$, where γ is sufficiently small, result in acceptable tracking quality. It has been shown in Subsections 3.2.4 and 3.3.3 that $\gamma \leq 0.4$ generally leads to good behavior. Clearly, the smaller γ, the better the quality of tracking. Nevertheless, some amount of quality degradation always occurs and, as mentioned in Subsection 3.2.4, can be due to either dynamic lagging or excessive oscillations. To prevent the latter, it is necessary to amend the admissible domain with a specification on I_3.

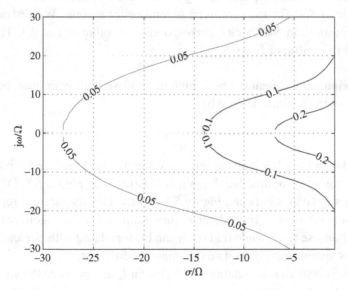

Figure 5.2. Level curves of I_2.

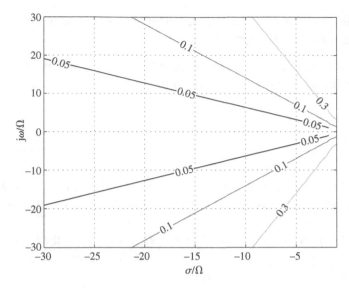

Figure 5.3. Level curves of I_3.

Admissible domains from the point of view of I_3: Figure 5.3 presents the level curves of I_3 in the above normalized coordinates. Since these level curves are almost radial straight lines, it follows that, as the damping ratio ζ of the closed loop poles decreases, the value of I_3 increases. Such an increase implies the appearance of oscillations in the output response (see, for instance, Figure 5.4, which shows the tracking quality for various values of I_3 with the same error standard deviation $\sigma_e = 0.1$). Therefore, it is of importance to determine the values of η in (5.22) that lead to acceptable oscillatory properties of tracking. This can be accomplished by ensuring that the sensitivity function, $S(s)$, does not amplify spectral components beyond the input bandwidth Ω. For that purpose, a design rule can be inferred from the magnitude characteristic of $S(s)$ for the prototype second order system. We restrict $S(s)$ to a peak of no more than 5dB, which corresponds to a value of $\zeta = 0.3$. This, in turn, corresponds to a value of $I_3 = 0.3$.

Complete admissible domain: The complete admissible domain now becomes the intersection of the regions defined by

$$I_2 \leq \gamma, I_3 \leq \eta, \tag{5.23}$$

where $\gamma \leq 0.4$ and $\eta \leq 0.3$. For the reference signal with $\Omega = 1$ and for $\gamma = 0.1$ and $\eta = 0.3$, the complete admissible domain is illustrated in Figure 5.5. Of immediate note are the similarities between Figure 5.5 and the classical desired region for the tracking of step references. Indeed, the requirement on I_2 is analogous to the classical requirement on rise time, while that on I_3 can be correlated with percent overshoot. Nevertheless, quantitatively the two domains are different.

Figure 5.6 illustrates the relationship between I_2 and σ_e when the standard deviation of the reference signal $\sigma_r = 1$ and $\zeta = 1$. Clearly, for $I_2 < 0.25$ this relationship is

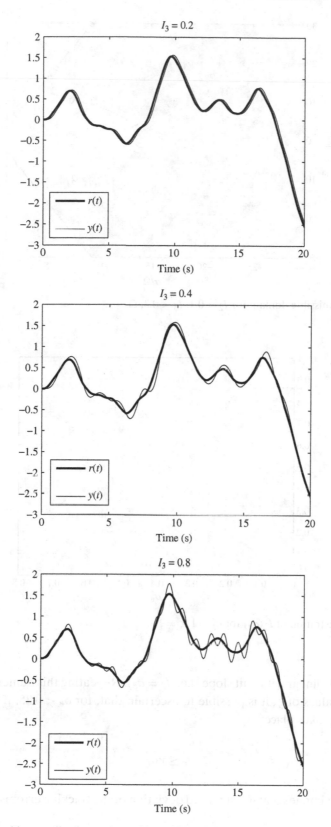

Figure 5.4. Tracking quality for various values of I_3, with $\sigma_e = 0.1$.

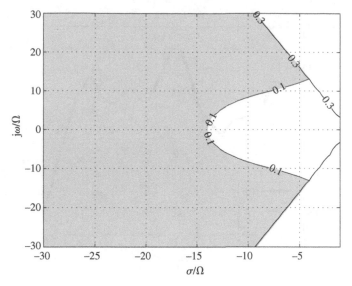

Figure 5.5. Admissible domain for $I_2 < 0.1$, $I_3 < 0.3$, $\Omega = 1$.

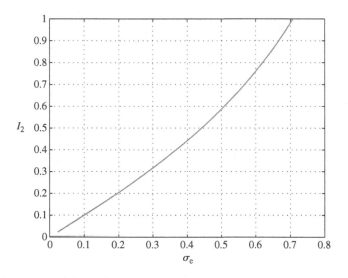

Figure 5.6. Illustration of $I_2(\sigma_e)$ for $\sigma_r = 1$, $\zeta = 1$.

approximately linear with unit slope (i.e., $I_2 = \sigma_e$). Repeating this numerical analysis for various values of ζ, it is possible to ascertain that, for $\sigma_e < 0.25$, if $I_2 = \gamma$, then the following takes place:

$$\sigma_e \leq \gamma \sigma_r. \tag{5.24}$$

Hence, $I_2 \leq \gamma$ implies that the standard deviation of the tracking error is at most $\gamma \sigma_r$.

Note that the above admissible domain has been obtained under the assumption that $\sigma_r = 1$. In general, however, σ_r may take arbitrary values. Clearly, due to linearity, the quality of tracking does not change relative to the magnitude of σ_r. Hence, the admissible domains constructed above remain valid for any σ_r.

5.1.3 Higher Order Systems

In classical controller design, it is often sufficient to consider a small number of dominant poles (and zeros) as a low order approximation to the full system dynamics. Moreover, in many of these cases, it is sufficient to consider a single pair of dominant poles, yielding an approximation by a prototype second order system. Generally, a valid low order approximation is constructed by attempting to satisfy the following condition as closely as possible:

$$\frac{|S_H(j\omega)|^2}{|S_L(j\omega)|^2} \cong 1, \forall \omega > 0. \tag{5.25}$$

Here, $S_H(s)$ denotes the actual high order sensitivity function, while $S_L(s)$ is the low order approximation.

Under the assumption that (5.25) holds, it follows directly from (5.8) that

$$\frac{RS_H(\Omega)}{RS_L(\Omega)} \cong 1, \forall \Omega, \tag{5.26}$$

where $RS_H(\Omega)$ and $RS_L(\Omega)$ are the high and low order random sensitivity functions obtained from $S_H(s)$ and $S_L(s)$, respectively. Consequently, the indicators I_2 and I_3 are roughly equal for the high and low order models, and we conclude that the high order system exhibits a quality of random tracking that is similar to that of its low order approximation.

5.1.4 Application: Hard Disk Servo Design

Consider the problem of hard disk servo control, where the objective is to maintain the disk head above a circular track. Due to various sources of irregularity, these tracks tend to deviate from a perfect circle, and the deviations may be modeled as bandlimited Gaussian processes. The resulting problem is one of tracking a random reference, and we may use the admissible domains derived above to design an appropriate controller.

The following performance specifications for this problem are considered (see Subsections 3.2.5 and 3.3.4):

$$\sigma_e \leq 0.06\sigma_r, \forall \Omega \leq 692\,\text{rad/sec}. \tag{5.27}$$

Thus, $\gamma = 0.06$, and from (5.24) we determine that to meet the specification it is required that $I_2 \leq 0.06$. Imposing the additional requirement that $I_3 \leq 0.3$, we obtain the admissible domain shown in Figure 5.7. Note that we have simply taken the

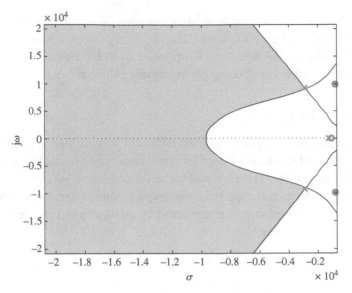

Figure 5.7. Admissible domain for the hard disk drive and pole-zero map of the closed loop system.

normalized admissible domain from Figure 5.2 and scaled the coordinates by $\Omega = 692$. Based on this domain, we design a controller such that a pair of dominant poles is placed at:

$$s_1, s_2 = (-2.83 \pm j9.06) \times 10^3,$$

which lie approximately at the intersection of the $I_2 = 0.06$ and $I_3 = 0.3$ level curves. The model of a disk servo is the same as in Subsection 3.2.5, that is,

$$P(s) = \frac{4.382 \times 10^{10}s + 4.382 \times 10^{15}}{s^2 \left(s^2 + 1.596 \times 10^3 s + 9.763 \times 10^7\right)}. \tag{5.28}$$

By employing a standard root locus based control design, using the above admissible domain, the following controller is obtained:

$$C(s) = \frac{K(s + 1058)\left(s^2 + 1596s + 9.763 \times 10^7\right)}{\left(s^2 + 3.719 \times 10^4 s + 5.804 \times 10^8\right)^2}, \tag{5.29}$$

where

$$K = 5.7214 \times 10^5. \tag{5.30}$$

Applying this controller to $P(s)$, results in the closed loop pole-zero configuration shown in Figure 5.7. Note that this controller leads to a number of approximate stable pole-zero cancellations and, consequently, yields a pair of dominant poles at the desired locations. The resulting tracking error is $\sigma_e = 0.0425\sigma_r$, which satisfies the design specifications given in (5.27). Figure 5.8 illustrates the quality of tracking in the time domain, where the output completely overlays the reference signal.

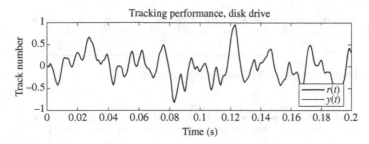

Figure 5.8. Tracking performance of the hard disk drive system.

5.2 Saturated Root Locus

5.2.1 Scenario

Consider the SISO tracking system of Figure 5.9. Here, $P(s)$ is the plant, $KC(s)$ is the controller, $K > 0$, and $F_\Omega(s)$ is a coloring filter with 3dB bandwidth Ω, which generates the reference r from standard white noise w_r; the signals y and u are, respectively, the system output and input to the saturation element defined as

$$\text{sat}_\alpha(u) = \begin{cases} \alpha, & u > +\alpha, \\ u, & -\alpha \leq u \leq \alpha, \\ -\alpha, & u < -\alpha. \end{cases} \tag{5.31}$$

As described in Chapter 2, stochastic linearization is used to study the system, whereby the saturation element is replaced by a gain, $N(K)$, defined as

$$N(K) = \text{erf}\left(\frac{\alpha}{\sqrt{2}\left\| \frac{F_\Omega(s)KC(s)}{1+N(K)KC(s)P(s)} \right\|_2} \right). \tag{5.32}$$

Note that, in this case, the stochastically linearized gain is expressed as a function of the free parameter K. The value of $N(K)$ can be calculated from (5.32) using Bisection Algorithm 2.1.

The locus traced by the closed loop poles of the quasilinear system of Figure 5.10 (referred to as *S-poles*) is called the *saturated root locus*, or *S-root locus* (SRL). It is the object of study in this section.

Denote the equivalent gain of the quasilinear system as

$$K_e(K) := KN(K). \tag{5.33}$$

Clearly, from (5.32), $K_e(K)$ can be obtained from the equation

$$K_e(K) = K\text{erf}\left(\frac{\alpha}{\sqrt{2}K\left\| \frac{F_\Omega(s)C(s)}{1+K_e(K)P(s)C(s)} \right\|_2} \right). \tag{5.34}$$

If $K_e(K) \to \infty$ as $K \to \infty$, the S-root locus may be the same as that of the feedback loop of Figure 5.9 in the absence of saturation (referred to as the *unsaturated* system).

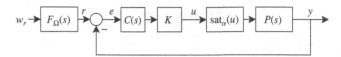

Figure 5.9. Closed loop system with saturating actuator.

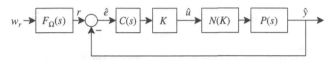

Figure 5.10. Equivalent quasilinear system.

If, however, $K_e(K) < \infty$ as $K \to \infty$, the S-root locus terminates at points prior to the open loop zeros. It turns out that the latter may indeed be the case, and in this chapter we show how these points, referred to as *S-termination points* can be calculated.

In addition, we investigate the relationship between the S-root locus and amplitude truncation of the reference signals. Clearly, this phenomenon does not arise in the unsaturated case. However, when the actuator is saturated, the trackable domain may be finite and, as a result, sufficiently large reference signals might be truncated. To indicate when this phenomenon takes place, we equip the S-root locus with the so-called amplitude *S-truncation points*, and provide methods for their calculation. As it turns out, both the S-termination and S-truncation points depend on all transfer functions in Figure 5.9, as well as on the level of saturation α.

5.2.2 Definitions

Definition 5.1. *The saturated closed loop poles (S-poles) of the nonlinear system of Figure 5.9 are the poles of the quasilinear system of Figure 5.10, that is, the poles of the transfer function from r to $\hat{y}$:*

$$T(s) = \frac{K_e(K)C(s)P(s)}{1 + K_e(K)C(s)P(s)}. \tag{5.35}$$

Definition 5.2. *The S-root locus is the path traced by the saturated closed loop poles when $K \in [0,\infty)$.*

Since $K_e(K)$ enters (5.35) as a usual gain, and

$$0 \le K_e(K) \le K,$$

the S-root locus is a proper or improper subset of the unsaturated root locus.

Define the auxilliary transfer function, $T_\gamma(s)$, as

$$T_\gamma(s) = \frac{F_\Omega(s)C(s)}{1 + \gamma P(s)C(s)}, \ \gamma \in \mathbf{R}^+. \tag{5.36}$$

It turns out that the properties of the S-root locus depend on the properties of $T_\gamma(s)$ and the solutions of (5.34). Specifically, we consider two cases: (i) when (5.36) is stable for all $\gamma > 0$, and (ii) when (5.36) is stable only on a finite interval $\gamma \in [0, \Gamma)$, $\Gamma < \infty$, and

$$\lim_{\gamma \to \Gamma} \|T_\gamma(s)\|_2 = \infty. \tag{5.37}$$

In addition, we consider cases where (5.34) has either a unique or multiple solutions.

5.2.3 S-Root Locus When $K_e(K)$ Is Unique

As in the unsaturated case, we are interested in the points of origin and termination of the S-root locus. The points of origin clearly remain the same as in the unsaturated case. The points of termination, however, may not be the same because, as it turns out, $K_e(K)$ may not tend to infinity as K increases. To discriminate between these two cases, consider the following equation

$$\beta - \left\| \frac{F_\Omega(s) C(s)}{1 + \left(\frac{\alpha \sqrt{2/\pi}}{\beta} \right) P(s) C(s)} \right\|_2 = 0, \tag{5.38}$$

for the unknown β. Note that, while (5.38) is always satisfied by $\beta = 0$, it may admit nonzero solutions as well. However, the uniqueness of $K_e(K)$ implies the uniqueness of a nonzero solution of (5.38). Indeed, the following holds:

Lemma 5.1. *If $K_e(K)$ is unique for all K, then*

  (i) *$K_e(K)$ is continuous and strictly monotonically increasing.*
  (ii) *Equation (5.38) admits at most one positive solution $\beta = \beta^* > 0$.*

Proof. See Section 8.4.

Theorem 5.1. *Assume that $T_\gamma(s)$ is asymptotically stable for all $\gamma > 0$ and (5.34) admits a unique solution for all $K > 0$. Then,*

$$\text{(i)} \quad \lim_{K \to \infty} K_e(K) = \frac{\alpha \sqrt{2/\pi}}{\beta^*} < \infty \tag{5.39}$$

if and only if (5.38) admits a unique solution $\beta = \beta^* > 0$;

$$\text{(ii)} \quad \lim_{K \to \infty} K_e(K) = \infty \tag{5.40}$$

if and only if $\beta = 0$ is the only real solution of (5.38).

Proof. See Section 8.4.

Theorem 5.2. *Assume that $T_\gamma(s)$ is asymptotically stable only for $\gamma \in [0, \Gamma)$, $\Gamma < \infty$, (5.37) holds, and (5.34) admits a unique solution for all $K > 0$. Then,*

$$\lim_{K \to \infty} K_e(K) = \frac{\alpha \sqrt{2/\pi}}{\beta^*} < \Gamma, \tag{5.41}$$

where $\beta = \beta^* > 0$ is the unique positive solution of (5.38).

Proof. See Section 8.4.

Note that (5.41) implies that, under the conditions of Theorem 5.2, the S-root locus can never enter the right-half plane. Thus, *a closed loop LPNI system with saturating actuator cannot be destabilized by an arbitrarily large gain* even if the relative degree of the plant is greater than two or if it has non-minimum phase zeroes.

As it follows from Theorems 5.1 and 5.2, in the limit as $K \to \infty$, the quasilinear system of Figure 5.10 has a closed loop transfer function given by

$$T_{ter}(s) = \frac{\kappa C(s)P(s)}{1 + \kappa C(s)P(s)}, \tag{5.42}$$

where

$$\kappa = \lim_{K \to \infty} K_e(K). \tag{5.43}$$

Definition 5.3. *The S-termination points of the S-root locus are the poles of $T_{ter}(s)$.*

Thus, as $K \to \infty$, the S-poles, defined by (5.35), travel monotically along the S-root locus from the open loop poles to the S-termination points defined by (5.42) and (5.43). If $\kappa = \infty$, then the S-termination points coincide with the open loop zeros; otherwise, the S-root locus terminates prematurely.

Example 5.1. Consider the system of Figure 5.9 with

$$P(s) = \frac{s + 15}{s(s + 2.5)}, \ \alpha = 0.16, \tag{5.44}$$

and $C(s), \Omega$, and $F_\Omega(s)$ given by

$$C(s) = 1, \ \Omega = 1 \, \text{rad/sec},$$

$$F_\Omega(s) = \sqrt{\frac{3}{\Omega}} \left(\frac{\Omega^3}{s^3 + 2\Omega s^2 + 2\Omega^2 s + \Omega^3} \right). \tag{5.45}$$

It is straightforward to verify that, for this system, $T_\gamma(s)$ is asymptotically stable for all $\gamma > 0$ and (5.34) admits a unique solution for $K > 0$, that is, the conditions of Theorem 5.1 are satisfied. Since (5.38) admits a positive solution $\beta = 2.58$, (5.39) and (5.43) result in $\kappa = 0.33$. Figure 5.11 shows $K_e(K)$ as a function of K. When K is small, we see that $K_e(K) \approx K$, since the actuator does not saturate. Clearly, as K increases, $K_e(K)$ indeed tends to the limit κ.

Figure 5.12 shows both the unsaturated and S-root loci of the system under consideration. The termination points are at $s = -1.4150 \pm 1.7169i$. Here, and in all subsequent figures, the S-termination points are indicated by small squares. The shaded area of Figure 5.12 represents the admissible domain for a high quality of tracking derived in Section 5.1. Clearly, the S-root locus never enters the admissible domain, and the achievable tracking quality is limited by the S-termination points.

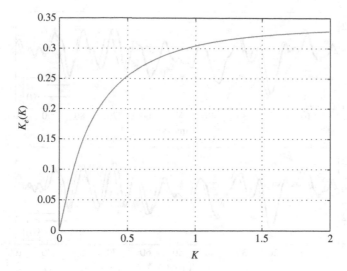

Figure 5.11. $K_e(K)$ for Example 5.1 with $\alpha = 0.16$.

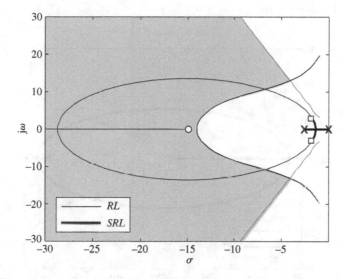

Figure 5.12. Unsaturated and S-root loci for Example 5.1 with $\alpha = 0.16$.

Figure 5.13 shows the output of the nonlinear and quasilinear systems when $K = 150$ (that is, when the saturated closed loop poles are located close to the S-termination points). Clearly, the tracking quality of the stochastically linearized system is poor due to dynamic lag. As predicted, the same is true for the original nonlinear system.

To improve the tracking performance, we increase α to 0.3 (here $\beta = 0$ is the only solution of (5.38), and, thus, the termination points are at the open loop zeroes). As illustrated in Figure 5.14, this causes the S-root locus to enter the admissible domain, and hence, choosing K large enough results in a high quality of tracking. This is verified in Figure 5.15, where we see that the quality of tracking is good for both the stochastically linearized and original nonlinear systems.

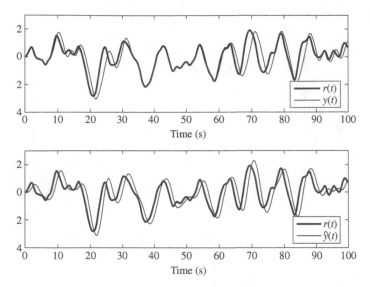

Figure 5.13. Tracking quality with S-poles located near the S-termination points ($K = 150$) for Example 5.1 with $\alpha = 0.16$.

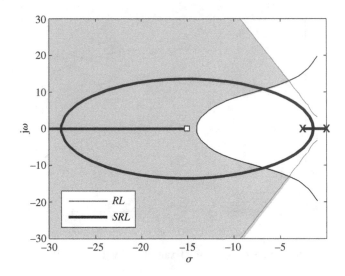

Figure 5.14. Unsaturated and S-root loci for Example 5.1 with $\alpha = 0.3$.

Example 5.2. Consider the system of Figure 5.9 with

$$P(s) = \frac{1}{s(s+1)(s+2)}, \ \alpha = 1, \tag{5.46}$$

and $C(s), \Omega$, and $F_\Omega(s)$ as defined in (5.45). It is easily verified that $T_\gamma(s)$ is stable only for $\gamma \in [0, 5.96)$ and (5.34) has a unique solution for $K > 0$, that is, the conditions of Theorem 5.2 are satisfied. Noting that $\beta = 1.1$ is the solution to (5.38), it follows from (5.41) and (5.43) that $\kappa = 0.722$. The resulting S-root locus, illustrated in Figure 5.16,

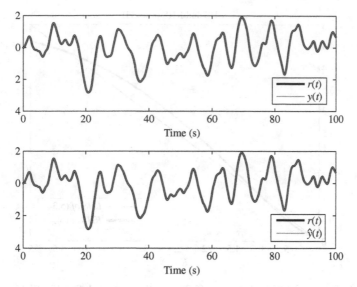

Figure 5.15. Tracking quality with poles located near the S-termination points $(K = 150)$ for Example 5.1 with $\alpha = 0.3$.

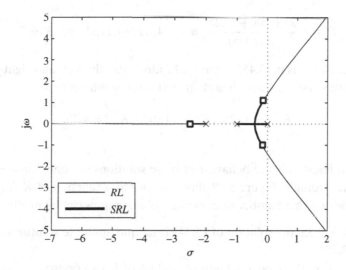

Figure 5.16. Unsaturated and S-root loci for Example 5.2.

never enters the right-half plane (the termination points are at $s = -2.2551, -0.3725 \pm 0.4260i$).

5.2.4 S-Root Locus When $K_e(K)$ Is Nonunique: Motivating Example

In some cases, (5.34) admits multiple solutions. To illustrate the complexities that arise in this situation we use the following motivating example.

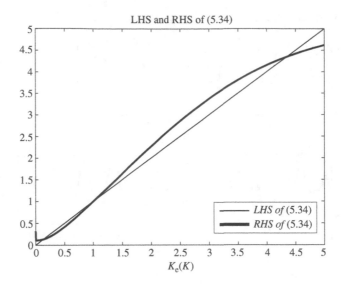

Figure 5.17. Left- and right-hand sides of (5.34) for the system defined in (5.47).

Consider the system of Figure 5.9 defined by

$$P(s) = \frac{s^2 + 10s + 27.25}{s^2(s+5)}, \ \alpha = 0.4, \ \Omega = 1.1 \, \text{rad/sec}, \ K = 5, \qquad (5.47)$$

with $C(s)$ and $F_\Omega(s)$ as in (5.45). Figure 5.17 illustrates the left- and right-hand sides of (5.34) for this system, where it is clear that three solutions exist:

$$K_e^{(1)} = 0.1283, K_e^{(2)} = 1.0153, K_e^{(3)} = 4.35. \qquad (5.48)$$

We are interested in the behavior of these solutions not only for $K = 5$, but for all $K > 0$. Accordingly, Figure 5.18 shows $K_e(K)$ as a function of K for the above example, from which a number of important observations can be made:

- For $K < 2.55$, the solution of (5.34) is unique; multiple solutions exist for all $K \geq 2.55$.
- As $K \to \infty$, three possible limiting values of $K_e(K)$ occur:

$$\kappa_1 = 0.1285, \ \kappa_2 = 1.0154, \ \kappa_3 = \infty. \qquad (5.49)$$

  Thus, the S-root locus has three different sets of S-termination points.
- At $K = 2.55$, the S-root locus has an *S-origination point*, corresponding to the appearance of two additional solutions.
- The range of $K_e(K)$ excludes the open interval between κ_1 and κ_2, that is, $(0.1285, 1.0154)$.

The branches of Figure 5.18 imply the following phenomenology of S-poles behavior for the system of Figure 5.9. For $K < 2.55$, there exists a unique set of

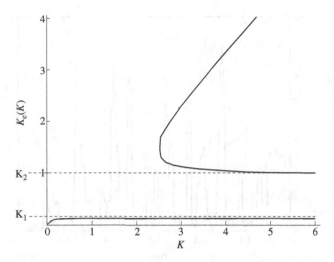

Figure 5.18. $K_e(K)$ as a function of K for the system defined in (5.47).

three S-poles, and the closed loop behaves accordingly. For $K > 2.55$, there are three sets of S-poles, and due to the stochastic nature, the system behavior can be characterized as jumping from one of these sets to another. It could be argued that the set of S-poles defined by the negative slope branch of Figure 5.18 does not correspond to a sustainable steady state, and therefore, the jumping phenomenon occurs between the poles defined by the positive slope branches of Figure 5.18.

To substantiate this phenomenology, consider the system of Figure 5.9 defined by (5.47). As mentioned before, this system leads to three values of $K_e(K)$ (defined in (5.48)) and, therefore, to three corresponding values of the steady state standard deviation of error:

$$\sigma_{\hat{e}}^{(1)} = 2.49, \; \sigma_{\hat{e}}^{(2)} = 0.31, \; \sigma_{\hat{e}}^{(3)} = 0.053. \tag{5.50}$$

Since the value of $\sigma_{\hat{e}}^{(2)}$ corresponds to $K_e(K)$ on the negative slope branch of Figure 5.18, it is expected that the closed loop system exhibits jumping between $\sigma_{\hat{e}}^{(1)}$ and $\sigma_{\hat{e}}^{(3)}$. To illustrate this behavior, we simulate the system at hand for 10,000 seconds and examine the standard deviation over a 100 second moving window. The results are shown in Figure 5.19. As one can see, the "simulated standard deviation," $\tilde{\sigma}_{\hat{e}}$, jumps roughly between the values of 0.05 and 2.5. This corroborates the described phenomenology.

These features give rise to the S-root locus illustrated in Figure 5.20, where the arrows indicate the movement of the S-poles as K increases. The S-termination points are denoted, as before, by the squares, while the S-origination points are indicated by strikes orthogonal to the root locus. The termination points are at $s = -5.0115, -5.0885, -0.0585 \pm 0.8338i, -0.4634 \pm 2.2853i$; the S-origination points are at $s = -5.1209, -0.6395 \pm 2.6535i$. The parts of the S-root locus corresponding to the jumping phenomenon are indicated by broken lines. The insets of Figure 5.20 are intended to magnify the S-root locus in the vicinity of the S-termination points.

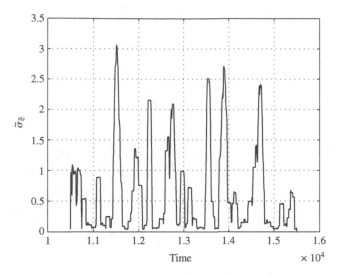

Figure 5.19. Illustration of jumping phenomenon of system defined in (5.47).

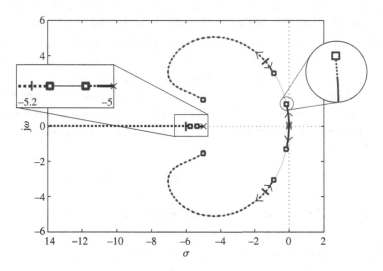

Figure 5.20. S-root locus for the system defined in (5.47).

Note that the S-root locus does not contain the parts of the unsaturated root locus corresponding to $K_e(K) \in (0.1283, 1.0153)$. Thus, the S-root locus for this system differs from the unsaturated case in the following:

- It contains areas of uncertain behavior due to the jumping phenomenon.
- It is only a proper subset of the unsaturated root locus, missing an intermediate portion.

5.2.5 S-Root Locus When $K_e(K)$ Is Nonunique: General Analysis

It is, of course, desirable to formulate general results pertaining to the multiple solutions case. Although the complexities outlined above lead to analytical difficulties, some useful deductions can be made.

Theorem 5.3. *Assume that* (5.34) *has multiple solutions, and the only solution of* (5.38) *is* $\beta = 0$. *Then:*

(i) *The S-root locus coincides with the unsaturated root locus, parameterized by* $K_e(K)$ *rather than* K.
(ii) *There exists an S-origination point, implying that there is a range of* K *for which the jumping phenomenon takes place.*

Proof. See Section 8.4.

As an illustration of Theorem 5.3, consider the system of (5.47), with $\alpha = 0.53$. In this case, (5.38) admits a unique solution $\beta = 0$, but $K_e(K)$ is non-unique for some K (see Figure 5.21). The resulting S-root locus is shown in Figure 5.22.

Theorem 5.4. *Assume that* (5.34) *has multiple solutions, and, along with* $\beta_0 = 0$, (5.38) *admits solutions* $\beta = \beta_i > 0$, $i = 1,2,...,n$. *Then:*

(i) *Each* β_i *corresponds to an S-termination point defined by*

$$\kappa_i = \frac{\alpha\sqrt{2/\pi}}{\beta_i}, \tag{5.51}$$

and (5.42).
(ii) *If* n *is even, then the solution* $\beta_0 = 0$ *also gives a termination point, defined by* $\kappa_0 = \infty$.

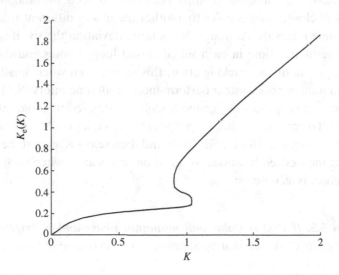

Figure 5.21. $K_e(K)$ as K tends to infinity for the system defined in (5.47) with $\alpha = 0.53$.

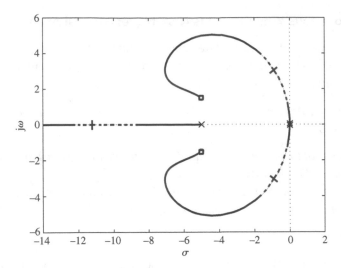

Figure 5.22. S-root locus for the system defined in (5.47) with $\alpha = 0.53$.

(iii) Assume that (5.38) admits at least one simple positive solution. Then the S-root locus is a proper subset of the unsaturated root locus.

Proof. See Section 8.4.

The motivating example (5.47) can be viewed as an illustration of Theorem 5.4.

5.2.6 Approach to Controller Design to Avoid Nonunique $K_e(K)$

As discussed above, a closed loop system with a controller $C(s)$ that leads to multiple solutions of (5.34), exhibits quite a complex behavior. Indeed, the random "jumping" from one set of closed loop S-poles to another, results in different tracking errors on different time intervals. Although, using large deviation theory, it is possible to evaluate the residence time in each set of closed loop S-poles and then estimate the "average" value of the tracking error, this value, even when small, would not necessarily guarantee good system performance for all time intervals.

Therefore, we propose a design methodology intended to avoid multiple solutions of (5.34). To accomplish this, we propose to select, if possible, a controller $C(s)$ that results in a unique $K_e(K)$ for all $K > 0$, and then select K so that the closed loop S-poles are at the desired locations. A question arises as to when such a $C(s)$ does exist. The answer is as follows:

Theorem 5.5. *If $P(s)$ is stable and minimum phase and $F_\Omega(s)/P(s)$ is strictly proper, there exists $C(s)$ such that the solution of (5.34) is unique for all $K > 0$.*

Proof. See Section 8.4.

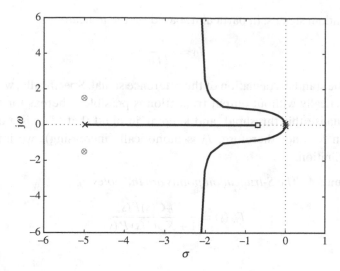

Figure 5.23. S-root locus for the system defined in (5.47) with $C(s)$ as in (5.52).

In the example of system (5.47), selecting

$$C(s) = \frac{s+0.7}{s^2 + 10s + 27.25},$$

(5.52)

results in a unique $K_e(K)$ (with $\kappa = \infty$, as per Theorem 5.1). The S-root locus for this case is illustrated in Figure 5.23.

5.2.7 S-Root Locus and Amplitude Truncation

In the previous sections we have used the S-root locus to characterize the dynamics of systems with saturating actuators. However, the performance may be poor not only due to the location of S-poles, but also due to output truncation by the saturation. Accordingly, in this section we introduce and compute S-truncation points, which characterize the region of the S-root locus where truncation does not occur. To accomplish this, we use the notion of trackable domain introduced in Section 3.1.

The trackable domain, TD, is defined by the magnitude of the largest step input that can be tracked in the presence of saturation. For the system of Figure 5.9, it is given by

$$|TD| = \left| \frac{1}{KC(0)} + P(0) \right| \alpha,$$

(5.53)

where $C(0)$ and $P(0)$ are the d.c. gains of the controller and plant, respectively. Clearly, the trackable domain is infinite when $P(s)$ has at least one pole at the origin; otherwise, it is finite, assuming that $C(0) \neq 0$. Although (5.53) is based on step signals, it has been shown in Section 3.2 that TD can also be used to characterize the tracking quality of random signals, in particular, based on the indicator

$$I_0 = \frac{\sigma_r}{|TD|}, \tag{5.54}$$

where σ_r is the standard deviation of the reference signal. Specifically, when $I_0 < 0.4$, tracking practically without output truncation is possible, whereas for $I_0 > 0.4$ it is not. To formalize this threshold, and keeping in mind that $|TD|$ is monotonically decreasing in K (and, therefore, I_0 is monotically increasing), we introduce the following definition.

Definition 5.4. *The S-truncation points are the poles of*

$$T_{tr}(s) = \frac{K_{tr}C(s)P(s)}{1 + K_{tr}C(s)P(s)}, \tag{5.55}$$

where

$$K_{tr} = K_e\left(K_{I_0}\right) \tag{5.56}$$

and

$$K_{I_0} = \min\{K > 0 : I_0 = 0.4\}. \tag{5.57}$$

Note that when $P(s)$ has a pole at the origin, $I_0 = 0$, and the S-root locus has no S-truncation points. To illustrate the utility of S-truncation points, we use the following example:

Example 5.3. Consider the system of Figure 5.9 defined by

$$P(s) = \frac{s+20}{(s+15)(s+0.5)}, C(s) = 1, \alpha = 0.8, \Omega = 1,$$

and $F_\Omega(s)$ as in (5.45). As obtained from Theorem 5.1, the limiting gain of this system is $\kappa = 20.015$ (the S-termination points are obtained through (5.42)). Figure 5.24 shows I_0 for this system as a function of K, from which we determine that $K_{I_0} = 2.2$. The truncation gain can then be evaluated as $K_{tr} = K_e(2.2) = 2.15$. The resulting S-root locus is given in Figure 5.25, where the S-termination and S-truncation points are denoted by the white and black squares, respectively (the S-termination points are at $s = -17.7575 \pm 9.6162i$; the S-truncation points are at $s = -3.5924, -14.057$). Although the S-root locus enters the shaded region for high-quality dynamic tracking, the position of the truncation points limit the achievable performance. Figure 5.26 illustrates the output response of the system when the control gain is $K = 20 > K_{I_0}$. The saturated closed loop poles are located at $-16 \pm j9.5$ (i.e., in the admissible domain), but beyond the truncation points. This leads to good dynamic tracking, but with a clipped response. Clearly, this can be remedied by increasing α, as illustrated in Figure 5.27, which shows the S-root locus when $\alpha = 0.92$ (the S-termination points are at $s = -20, -\infty$; the S-truncation points are at $s = -17.5800 + \pm 9.5731i$). Figure 5.28 shows the corresponding plot for the same location of S-poles as before. Clearly the clipping practically does not occur (the output overlays the reference).

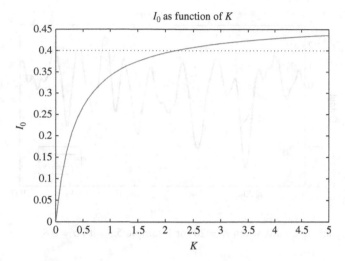

Figure 5.24. I_0 as a function of K for Example 5.3.

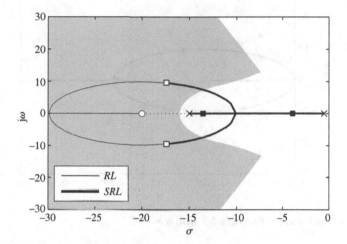

Figure 5.25. Unsaturated and S-root loci for Example 5.3 with $\alpha = 0.8$.

In conclusion, note that if S-truncation points exist, then they occur prior to the S-termination points (since the latter correspond to $K = \infty$).

5.2.8 Calibration of the S-Root Locus

Let s^* be an arbitrary point on the S-root locus, that is,

$$1 + K_e(K) C(s^*) P(s^*) = 0, \qquad (5.58)$$

where

$$0 \le K_e(K) < \kappa. \qquad (5.59)$$

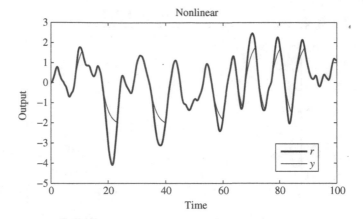

Figure 5.26. Tracking quality when $K = 20$ for Example 5.3 with $\alpha = 0.8$.

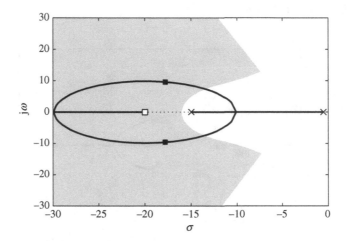

Figure 5.27. S-root locus for Example 5.3 with $\alpha = 0.92$.

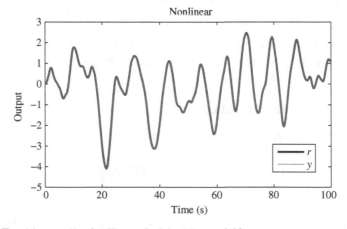

Figure 5.28. Tracking quality for Example 5.3 with $\alpha = 0.92$.

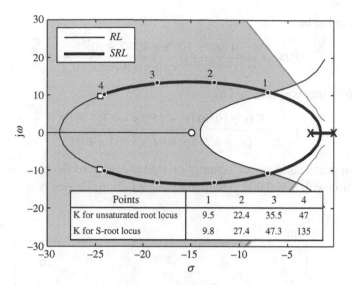

Figure 5.29. Difference in calibration between unsaturated and S-root locus.

To calibrate the S-root locus means to find the particular K such that (5.58) is satisfied (i.e., the S-poles are located at s^*). This is accomplished using the following corollary of Theorem 5.1.

Corollary 5.1. *For arbitrary s^* on the S-root locus, there exists a unique $K^* > 0$ such that $K = K^*$ satisfies (5.58). Moreover, K^* is the unique solution of*

$$K_e = K^* \mathrm{erf}\left(\frac{\alpha}{\sqrt{2}K^* \left\| \frac{F_{\Omega}(s)C(s)}{1+K_e P(s)C(s)} \right\|_2} \right), \qquad (5.60)$$

where

$$K_e = \frac{1}{|C(s^*)P(s^*)|}. \qquad (5.61)$$

Proof. See Section 8.4.

Note that (5.60) can be solved by a standard bisection algorithm. Figure 5.29 illustrates the differences in calibration gains between an unsaturated and S-root locus (using the system of Example 5.1).

5.2.9 Application: LPNI Hard Disk Servo Design

We consider the hard disk drive servo problem of Subsection 5.1.4, where the control objective is to maintain the disk head above a circular track that exhibits random irregularities, which can be modeled as a bandlimited noise. The model for the plant was considered to be

$$P(s) = \frac{4.382 \times 10^{10}s + 4.382 \times 10^{15}}{s^2\left(s^2 + 1.596 \times 10^3 s + 9.763 \times 10^7\right)}, \tag{5.62}$$

while the controller was selected as

$$C(s) = \frac{K(s + 1058)\left(s^2 + 1596s + 9.763 \times 10^7\right)}{\left(s^2 + 3.719 \times 10^4 s + 5.804 \times 10^8\right)^2}. \tag{5.63}$$

Here, we impose the additional assumption that the input to the plant is constrained by a saturation with $\alpha = 0.006$. The bandwidth of the reference is 692 rad/sec, and

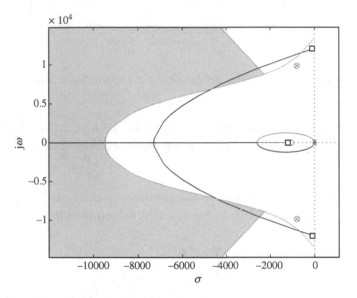

Figure 5.30. S-root locus for hard disk drive example.

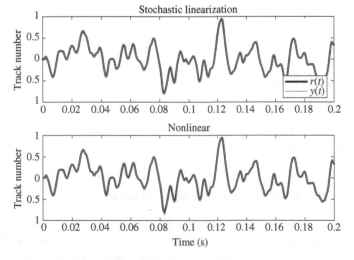

Figure 5.31. Tracking performance, hard disk drive example.

$F_\Omega(s)$ is as defined in (5.45). For this system, the conditions of Theorem 5.2 hold, and it results in

$$\kappa = 1.0509 \times 10^6.$$

The corresponding S-root locus is shown in Figure 5.30, where the admissible domain is indicated by the shaded region. Note that since $P(s)$ contains poles at the origin, the S-root locus does not have S-truncation points. To meet the performance specification, we select K so that

$$K_e(K) = 5.7214 \times 10^5, \tag{5.64}$$

which results in a pair of dominant S-poles located at $(-2.83 \pm j9.06) \times 10^3$, that is, in the admissible domain. Using Corollary 5.1 with (5.64) yields:

$$K = 6.208 \times 10^5.$$

Figure 5.31 illustrates the tracking quality for both the stochastically linearized and original nonlinear systems. As predicted, the nonlinear system exhibits a tracking quality similar to that of the stochastically linearized system, and achieves a standard deviation $\sigma_e = 0.047\sigma_r$. It is noteworthy that this performance matches that obtained by a more complicated nonlinear anti-windup design methodology.

5.3 Summary

- In the problem of random reference tracking in linear systems, the admissible domains of closed loop poles are determined by the quality indicators introduced in Chapter 3. It turns out that these domains are qualitatively similar to but quantitatively different from those for tracking deterministic references (i.e., defined by the settling time and the overshoot).
- The closed loop poles of LPNI systems are defined as the closed loop poles of the corresponding stochastically linearized systems. The locus traced by these poles, as the controller gain changes from 0 to ∞, is referred to as the root locus of an LPNI system.
- For LPNI systems with saturating actuator, the S-root locus originates at open loop poles and ends at the so-called termination points. The location of the termination points are calculated using a transcendental equation. If this equation has a positive solution, the S-root locus terminates prematurely – before it reaches the open loop zeros.
- In addition, the S-root loci are equipped with the so-called truncation points – beyond these points, amplitude truncation takes place. An equation for calculating these points is provided.
- If the quasilinear gain is not unique, the LPNI system has multiple sets of closed loop poles, each corresponding to a particular solution of the transcendental equation. In this case, the system exhibits a "jumping" pattern of behavior – defined by one set of the poles or another.

- Similar to the usual root locus, the S-root locus can be used for designing random reference tracking controllers for LPNI systems in the time domain.
- Root loci for LPNI systems with nonlinearities other than saturation can be constructed using the same approach as in the S-root locus case.
- The S-root locus and admissible domains provide a time domain solution of the Narrow Sense Design problem introduced in Section 1.2.

5.4 Problems

Problem 5.1. Consider the tracking system of Figure 5.1. Assume that

$$C(s)P(s) = \frac{\omega_n^2}{s(s+2\zeta\omega_n)} \tag{5.65}$$

and $F_\Omega(s)$ is the 3rd order Butterworth filter

$$F_\Omega(s) = \sqrt{\frac{3}{\Omega}} \left(\frac{\Omega^3}{s^3 + 2\Omega s^2 + 2\Omega^2 s + \Omega^3} \right) \tag{5.66}$$

driven by standard white noise. For $\Omega = 10$ rad/sec, construct the admissible domain for the following values of tracking quality indicators:

(a) $I_2 < 0.2, I_3 < 0.4$;
(b) $I_2 < 0.1, I_3 < 0.3$;
(c) $I_2 = 0.1, I_3 = 0.2$.

Problem 5.2. Repeat Problem 5.1 using the coloring filter

$$F_\Omega(s) = \frac{\Omega}{s+\Omega} \tag{5.67}$$

and comment on the difference of the results.

Problem 5.3. Consider the tracking system of Figure 5.1 with $C(s)P(s)$ and $F_\Omega(s)$ as in Problem 5.1. Assume that the reference bandwidth is $\Omega = 16$ rad/sec. For each of the following values of the damping ratio ζ, determine the range of ω_n so that the closed loop poles lie in the admissible domain defined by $I_2 < 0.2, I_3 < 0.4$:

(a) $\zeta = 0.3$;
(b) $\zeta = 0.707$;
(c) $\zeta = 1$.

Problem 5.4. Consider the tracking system of Figure 5.1 with

$$P(s) = \frac{1}{s(s+2)(s+4)}, \tag{5.68}$$

and $F_\Omega(s)$ as in (5.66). Assume that the reference bandwidth is $\Omega = 16$ rad/sec.

(a) Design a controller $C(s)$ so that the dominant closed loop poles lie in the admissible domain defined by $I_2 < 0.2, I_3 < 0.4$.

(b) By simulation, check if the resulting closed loop system exhibits the intended performance.

Problem 5.5. Consider the tracking system of Figure 5.1 with

$$P(s) = \frac{88.76}{s(s+21.53)(s+2.474)}.$$ (5.69)

Assume that the control objective is to track a reference signal of bandwidth $\Omega = 200$ rad/sec generated by the filter (5.66) driven by standard white noise.

(a) Find a specification in terms of I_2 and I_3 that ensures no more than 10% tracking error.
(b) Using the resulting admissible domain and the notion of dominant poles, design a controller that meets the specification.
(c) Now, assume that the voltage at the input of the plant is known to saturate with the limits $\alpha = \pm 0.5V$, that is, the system has the structure of Figure 5.9 with $K = 1$. Simulate the system and investigate the tracking performance.
(d) Explain the tracking performance by evaluating the closed loop S-poles and their proximity to the admissible domain of part (b).

Problem 5.6. Consider the tracking system of Figure 5.9 with the plant specified by

$$P(s) = \frac{s+10}{s(s+1)}.$$ (5.70)

Assume that the reference signal is generated by the coloring filter (5.66) with $\Omega = 10$ rad/sec driven by standard white noise. The actuator is the saturation $\text{sat}_\alpha(\cdot)$ and the controller is $C(s) = 1$.

(a) Compute the limiting gain κ and S-termination points for the following values of α:
    i. $\alpha = 0.1$,
    ii. $\alpha = 0.2$,
    iii. $\alpha = 1$.
(b) Draw the corresponding S-root loci.
(c) Does this tracking system exhibit amplitude truncation?

Problem 5.7. Consider the tracking system of Figure 5.9 with the plant specified by

$$P(s) = \frac{s+10}{(s+2)(s+1)}.$$ (5.71)

Assume that the reference signal is generated by the coloring filter (5.66) with $\Omega = 10$ rad/sec driven by standard white noise.

(a) Compute the S-termination and S-truncation points for
    i. $\alpha = 0.1$,
    ii. $\alpha = 0.2$,
    iii. $\alpha = 1$.

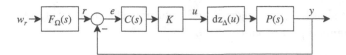

Figure 5.32. Problem 5.11.

(b) Draw the corresponding S-root loci.

Problem 5.8. Consider the tracking system of Figure 5.9 with the plant speci-fied by

$$P(s) = \frac{1}{s(s+1)}, \tag{5.72}$$

and reference signal given by the coloring filter (5.66) with $\Omega = 10$ rad/sec driven by standard white noise. Determine the S-termination points and draw the S-root locus for each of the following cases:

(a) $C(s) = 1, \alpha = 1$;
(b) $C(s) = 10, \alpha = 1$;
(c) $C(s) = 1, \alpha = 0.1$;
(d) Comment on the changes to the S-root locus due to changes in $C(s)$ and α.

Problem 5.9. Consider the tracking system of Figure 5.9 with

$$P(s) = \frac{s^2 + 10}{s(s^2 + 5)}, C(s) = 10, K = 1. \tag{5.73}$$

Assume that the reference signal is generated by the coloring filter (5.66) with $\Omega = 10$ rad/sec driven by standard white noise.

(a) Numerically determine the range of α for which (5.38) admits multiple solutions.
(b) Choose one such value of α and simulate the system to observe the bistable, that is, jumping behavior.

Problem 5.10. Consider the tracking system of Figure 5.9 with the plant

$$P(s) = \frac{88.76}{s(s+21.53)(s+2.474)}, \tag{5.74}$$

and proportional controller $C(s) = 1$. Assume that the reference signal is generated by the coloring filter (5.66) driven by standard white noise. Determine the minimum value of α such that a reference of bandwidth $\Omega = 20$ rad/sec can be tracked with high quality. Use the admissible domain $I_2 < 0.2$, $I_3 < 0.3$.

Problem 5.11. Consider the tracking system of Figure 5.32 where the actuator is a deadzone nonlinearity $dz_\Delta(u)$ defined by (2.23). Develop a method for constructing a "DZ-root locus." Specifically,

(a) Find expressions for the limiting gain.
(b) Find expressions for DZ-termination points.

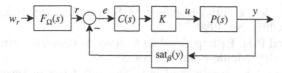

Figure 5.33. Problem 5.14.

(c) Plot the DZ-root locus for the system defined by

$$P(s) = \frac{1}{s(s+1)}, \quad C(s) = 1, \quad \Delta = 1.$$

Problem 5.12. Repeat Problem 5.11 for the case of actuator quantization defined by (2.35).

Problem 5.13. Repeat Problem 5.11 for the case of a quantized actuator with saturation given by (2.38).

Problem 5.14. Develop a method for constructing a "sensor S-root locus," which we define as S_s-root locus. For this purpose, consider the LPNI tracking system of Figure 5.33, where the actuator is linear and the sensor is the saturation $\mathrm{sat}_\beta(y)$. Note that in this case, the equivalent sensor gain is given by:

$$N_s(K) = \mathrm{erf} \left(\frac{\beta}{\sqrt{2} \left\| \frac{F_\Omega(s)P(s)C(s)K}{1+P(s)C(s)KN_s} \right\|_2} \right). \tag{5.75}$$

(a) Find an expression for the limiting gain

$$\kappa = \lim_{K \to \infty} KN_s(K). \tag{5.76}$$

(b) Find expressions for the S_s-termination points.
(c) Find, if any, expressions for the S_s-truncation points.
(d) Draw the S_s-root locus for

$$P(s) = \frac{1}{s(s+1)}, \quad C(s) = 1, \quad \beta = 0.5, \tag{5.77}$$

where $F_\Omega(s)$ is the coloring filter (5.66) with $\Omega = 10\,\mathrm{rad/sec}$ driven by standard white noise.

5.5 Annotated Bibliography

The classical notions of admissible domain and root locus for linear systems can be found in any undergraduate text on control, see, for instance, the following:

[5.1] B.C. Kuo, *Automatic Control Systems*, Fifth Edition, Prentice Hall, Englewood Cliffs, NJ, 1987

[5.2] K. Ogata, *Modern Control Engineering*, Second Edition, Prentice Hall, Englewood Cliffs, NJ, 1990

[5.3] R.C. Dorf and R.H. Bishop, *Modern Control Systems*, Eighth Edition, Addison-Wesley, Menlo Park, CA, 1998

[5.4] G.C. Goodwin, S.F. Graebe, and M.E. Salgado, *Control Systems Design*, Prentice Hall, Upper Saddle River, NJ, 2001

[5.5] G.F. Franklin, J.D. Powel, and A. Emami-Naeini, *Feedback Control of Dynamic Systems*, Fourth Edition, Prentice Hall, Englewood Cliffs, NJ, 2002

The material of Section 5.1 is based on the following:

[5.6] S. Ching, P.T. Kabamba, and S.M. Meerkov, "Admissible pole locations for tracking random references," *IEEE Transactions on Automatic Control*, Vol. 54, pp. 168–171, 2009

As mentioned before, a description of hard disk servo problem can be found in the following:

[5.7] T.B. Goh, Z. Li, and B.M. Chen, "Design and implementation of a hard disk servo system using robust and perfect tracking approach," *IEEE Transactions on Control Systems Technology*, Vol. 9, 221–233, 2001

The material of Section 5.2 is based on the following:

[5.8] S. Ching, P.T. Kabamba, and S.M. Meerkov, "Root locus for random reference tracking in systems with saturating actuators," *IEEE Transactions on Automatic Control*, Vol. 54, pp. 79–91, 2009

6 Design of Disturbance Rejection Controllers for LPNI Systems

Motivation: The LQR and LQG techniques are widely used for designing disturbance rejection controllers for linear systems. In this chapter, quasilinear versions of these techniques are developed and extended to simultaneous design of controllers and instrumentation, that is, actuators and sensors.

Overview: We begin with developing the SLQR/SLQG technique, where the "S" stands for "saturated" (Section 6.1). Then, we extend LQR/LQG to the so-called ILQR/ILQG (where the "I" stands for "instrumented"), which is intended for simultaneous design of controllers, sensors, and actuators (Section 6.2). For each of these methods, we show that the solution is given by the usual Lyapunov and Ricatti equations coupled with additional trancendental equations, which account for the Lagrange multipliers of the associated optimization problems and the quasilinear gains. Bisection algorithms for solving these coupled equations are provided.

6.1 Saturated LQR/LQG

6.1.1 Scenario

The LQR/LQG design methodology is one of the main state space techniques for designing linear controllers for linear plants, assuming that the actuators and sensors are also linear. Clearly, this methodology is not applicable to nonlinear systems. Is an extension to LPNI systems possible? A positive answer to this question is provided in this section for systems with saturating actuators. The resulting technique is referred to as SLQR/SLQG, where the "S" stands for "saturated." Extensions to other nonlinearities and to nonlinearities in actuators and sensors simultaneously are also possible using the approach presented here.

This approach can be outlined as follows: First, the equations of the closed loop LPNI system are quasilinearized using stochastic linearization. Then, the usual Lagrange multiplier optimization technique is applied to the resulting quasilinear system with a quadratic performance index. Since the gain of the quasilinear system depends on the standard deviation of the signals at the input of the nonlinearities (either actuators, or sensors, or both), the resulting synthesis relationships consist not only of the usual Lyapunov and Ricatti equations but also transcendental equations, which specify the quasilinear gain and the Lagrange multiplier associated with the minimization problem. Finally, the existence and uniqueness of solutions to the synthesis equations are analyzed, and an algorithm for computing these solutions is presented.

In addition to the above, we show that actuator saturation leads to performance limitations of optimal closed loop LPNI systems, similar to those of optimal linear systems with non-minimum phase plants. Thus, closed loop systems with saturating actuators cannot reject disturbances to arbitrary levels, even if the plant is minimum phase. A method for calculating the limit of achievable performance is provided. Also, we show that to obtain the optimal performance, the actuator must, in fact, saturate. Thus, design approaches based on saturation avoidance are not optimal.

Since the SLQR/SLQG controllers are intended to be used in the original closed loop LPNI system, one more problem is addressed: the problem of stability and performance *verification* of closed loop LPNI systems with SLQR/SLQG controllers. In this regard, we show that the property of (global) exponential stability of the quasilinear system is inherited by the closed loop LPNI system in a weaker, that is, semi-global, sense. Lastly, we indicate that the level of disturbance rejection of the closed loop LPNI system is, in most cases, close to that of the corresponding optimal quasilinear system.

6.1.2 Problem Formulation

Consider the system shown in Figure 6.1, where $P(s)$ is the plant, $C(s)$ is the controller, $sat_\alpha(u)$ and $A(s)$ are the nonlinearity and the dynamics of the actuator, and $F_i(s)$ and $H_i(s)$, $i = 1,2$ are coloring and weighting filters, respectively. Signals u, v, y, $y_m \in \mathbf{R}$ are the controller output, actuator output, plant output, and measured output, respectively, w_1, $w_2 \in \mathbf{R}$ are standard uncorrelated white noise processes, z_1, $z_2 \in \mathbf{R}$ are the controlled outputs.

The assumption that w_1 and w_2 are standard white noise processes (i.e., of unit intensity) does not restrict generality since it can be enforced by rescaling the plant

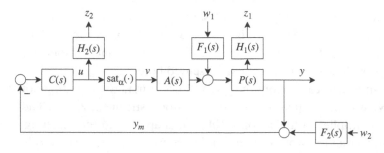

Figure 6.1. System model.

and the controller. The assumption that all signals are scalar is made to simplify the presentation. An extension to the vector case is discussed in Subsection 6.1.6.

Assume that the system, excluding the controller, has the state-space representation

$$
\begin{aligned}
\dot{x}_G &= Ax_G + B_1 w + B_2 \mathrm{sat}_\alpha(u), \\
z &= C_1 x_G + D_{12} u, \\
y_m &= C_2 x_G + D_{21} w,
\end{aligned}
\tag{6.1}
$$

where $x_G = [x_P^T \, x_A^T \, x_{F_1}^T \, x_{F_2}^T \, x_{H_1}^T \, x_{H_2}^T]^T \in \mathbb{R}^{n_{x_G}}$, $w = [w_1 \, w_2]^T$, and $z = [z_1 \, z_2]^T$. Let σ_z^2 denote the steady state variance of z.

The stochastically linearized version of (6.1) is

$$
\begin{aligned}
\dot{\hat{x}}_G &= A\hat{x}_G + B_1 w + B_2 N\hat{u}, \\
\hat{z} &= C_1 \hat{x}_G + D_{12}\hat{u}, \\
\hat{y}_m &= C_2 \hat{x}_G + D_{21} w, \\
N &= \mathrm{erf}\left(\frac{a}{\sqrt{2}\sigma_{\hat{u}}}\right),
\end{aligned}
\tag{6.2}
$$

where $\sigma_{\hat{u}}$ is the standard deviation of the controller output.

The main problem considered in this section is: *Given (6.2), find a controller $C(s)$, which stabilizes the stochastically linearized closed loop system when $w = 0$ and minimizes $\sigma_{\hat{z}}^2$ when $w \neq 0$.* This problem is referred to as SLQR in the case of state space feedback and SLQG in the case of dynamic output feedback.

Solutions of the SLQR and SLQG problems and corresponding verification problems are given in Subsections 6.1.3 and 6.1.4, respectively.

6.1.3 SLQR Theory

SLQR synthesis equations: Consider the open loop system

$$
\begin{aligned}
\dot{x}_G &= Ax_G + B_1 w + B_2 \mathrm{sat}_\alpha(u), \\
z &= C_1 x_G + D_{12} u
\end{aligned}
\tag{6.3}
$$

with the state feedback controller

$$
u = Kx_G,
\tag{6.4}
$$

and assume the following:

Assumption 6.1.
(i) (A, B_2) is stabilizable;
(ii) (C_1, A) is detectable;
(iii) $D_{12} = [0 \ \sqrt{\rho}]^T$, $\rho > 0$;
(iv) $D_{12}^T C_1 = 0$;
(v) *A has no eigenvalues in the open right-half plane.*

Assumptions (i)–(iv) are standard in the LQR theory. Assumption (v) is introduced to achieve semi-global stability of the closed loop system (6.3) and (6.4).

The closed loop system (6.3) and (6.4) is governed by

$$\dot{x}_G = Ax_G + B_2 \text{sat}_\alpha(Kx_G) + B_1 w,$$
$$z = (C_1 + D_{12}K)x_G. \tag{6.5}$$

Application of stochastic linearization to this system yields

$$\dot{\hat{x}}_G = (A + B_2 NK)\hat{x}_G + B_1 w,$$
$$\hat{z} = (C_1 + D_{12}K)\hat{x}_G, \tag{6.6}$$
$$N = \text{erf}\left(\frac{a}{\sqrt{2}\sigma_{\hat{u}}}\right),$$

Theorem 6.1. *Under Assumption 6.1, the SLQR problem*

$$\min_K \sigma_{\hat{z}}^2, \tag{6.7}$$

where the minimization is over all K such that $A + B_2 NK$ is Hurwitz, has a unique solution. The minimum value of the cost is

$$\min_K \sigma_{\hat{z}}^2 = \text{tr}\left(C_1 R C_1^T\right) + \rho \frac{N^2}{(\rho+\lambda)^2} B_2^T QRQB_2, \tag{6.8}$$

and a state feedback gain K that achieves this minimum is

$$K = -\frac{N}{\rho+\lambda} B_2^T Q, \tag{6.9}$$

where (Q,R,N,λ) is the unique solution of the following system of equations

$$A^T Q + QA - \frac{N^2}{\rho+\lambda} QB_2 B_2^T Q + C_1^T C_1 = 0, \tag{6.10}$$

$$(A - \frac{N^2}{\rho+\lambda} B_2 B_2^T Q)R + R(A - \frac{N^2}{\rho+\lambda} B_2 B_2^T Q)^T + B_1 B_1^T = 0, \tag{6.11}$$

$$\left(\frac{N^2}{\rho+\lambda}\right)^2 B_2^T QRQB_2 - \frac{N^2 \alpha^2}{2\left[\text{erf}^{-1}(N)\right]^2} = 0, \tag{6.12}$$

$$\lambda - \frac{\rho}{\frac{\sqrt{\pi}}{2}\frac{N}{\text{erf}^{-1}(N)} \exp\left(\left[\text{erf}^{-1}(N)\right]^2\right) - 1} = 0. \tag{6.13}$$

In addition, if R is nonsingular, then the state feedback gain K given by (6.9) is unique.

Proof. See Section 8.5.

In the proof of this theorem, it is shown that the left-hand side of (6.12), viewed as an implicit function of N, is continuous and changes sign between 0 and 1 only once. This suggests the following algorithm to compute the SLQR feedback gain.

SLQR Bisection Algorithm 6.1: Given a desired accuracy $\epsilon > 0$,

(a) Start with $N_1 = 0$ and $N_2 = 1$
(b) Let $N = (N_1 + N_2)/2$
(c) Calculate λ from (6.13)
(d) Solve the Riccati equation (6.10) for Q
(e) Solve the Lyapunov equation (6.11) for R
(f) Calculate the left-hand side of (6.12) and call it δ
(g) If $|N_2 - N_1| < \epsilon$, go to step (i)
(h) If $\delta < 0$, let $N_1 = N$, else let $N_2 = N$, and go to step (b)
(i) Calculate K from (6.9).

Corollary 6.1. *SLQR Bisection Algorithm 6.1 provides the solution of the SLQR problem with desired accuracy after a finite number of iterations.*

Proof. See Section 8.5.

Note that equations (6.8)–(6.13) reduce to the standard LQR equations when the saturation is removed. Indeed, letting α tend to infinity, it is easy to see that N and λ approach 1 and 0, respectively. Thus, SLQR is a proper extension of LQR.

SLQR performance limitations: It is well known in LQR theory that any arbitrarily small output variance can be attained if the plant is minimum phase. Below, it is shown that this is not true in the SLQR case, and limits for the best achievable performance are provided.

Theorem 6.2. *Let Assumption 6.1 hold, assume $C_1(sI - A)^{-1}B_1 \not\equiv 0$, and view ρ of Assumption 6.1(iii) as a parameter. Denote the first term in the optimal value of the cost expression (6.8) as $\gamma^2(\rho)$, that is,*

$$\gamma^2(\rho) = \operatorname{tr}\left(C_1 R(\rho) C_1^T\right). \tag{6.14}$$

Then, $\gamma^2(\rho)$ is an increasing function of ρ and

$$\lim_{\rho \to 0^+} \gamma^2(\rho) = \gamma_0^2 > 0. \tag{6.15}$$

Proof. See Section 8.5.

Thus, the variance of $\hat{z}_1$ cannot be made smaller than γ_0^2, and this value can be determined using SLQR Bisection Algorithm 6.1 with sufficiently small ρ.

Optimal level of saturation: It is of interest to investigate whether and to what extent the SLQR controller activates the saturation. This is quantified below.

Corollary 6.2. *The SLQR controller results in saturation activation given by*

$$P\{|\hat{u}| > 1\} = 1 - N, \tag{6.16}$$

where N with R, Q, and λ are the solution of (6.10)–(6.13).

Proof. See Section 8.5.

Thus, *to minimize $\sigma_{\tilde{z}}^2$, the actuator should experience saturation to the degree defined by (6.16)*.

Stability verification: Below, we analyze stability properties of the original system (6.3) with the SLQR controller (6.9).

Theorem 6.3. *Consider the undisturbed version of the closed loop system (6.3), (6.4):*

$$\dot{x}_G = Ax_G + B_2 \text{sat}_\alpha(u),$$
$$z = C_1 x_G + D_{12} u. \tag{6.17}$$

Let Assumption 6.1 hold, assume (Q, R, N, λ) is the solution of (6.10)–(6.13) and K is the SLQR gain (6.9). Then

(i) *$x_G = 0$ is the unique equilibrium point of (6.17), (6.9);*
(ii) *this equilibrium is exponentially stable;*
(iii) *a subset of its domain of attraction is given by*

$$\mathcal{X} = \{x_G \in \mathbf{R}^{n_{x_G}} : x_G^T(\varepsilon Q)x_G \leq \frac{4\alpha^2}{B_2^T(\varepsilon Q)B_2}\}, \tag{6.18}$$

where

$$\varepsilon = \frac{N^2}{\rho + \lambda}; \tag{6.19}$$

(iv) *if A is Hurwitz, all solutions of the closed loop system (6.17), (6.9) are bounded.*

Proof. See Section 8.5.

Note that since the origin of the undisturbed system (6.17), (6.9) is exponentially stable, it follows that the closed loop system (6.3), (6.9) is also small-signal finite-gain $\mathcal{L}_2$ stable.

Next, we show that by choosing an appropriate ρ, any set in $\mathbf{R}^{n_{x_G}}$ can be made a subset of the domain of attraction of (6.17), (6.9).

Theorem 6.4. *Under Assumption 6.1, for every bounded set $\mathcal{B}$, there exists a $\rho^* > 0$ such that, for all $\rho \geq \rho^*$, the closed loop system (6.17), (6.9) is exponentially stable and $\mathcal{B}$ is contained in its domain of attraction.*

Proof. See Section 8.5.

Thus, although the undisturbed version of the stochastically linearized closed loop system (6.6), (6.9) is globally exponentially stable, the real closed loop system (6.17), (6.9) inherits this property in a weaker sense, that is, in the sense of semi-global stability. Of course, to obtain a given domain of attraction, the performance index (and therefore, the controller) should be modified accordingly.

Note that since for a given controller, the closed loop system (6.17), (6.9) is only asymptotically stable, rather than globally asymptotically stable, additive disturbance $B_1 w$ (see equation (6.3)) may, in principle, force the system outside of the domain of attraction and, perhaps, even destabilize the closed loop behavior. To check whether this is possible, one may use the Popov stability criterion and determine if (6.17), (6.9) is in the Popov sector. If it is, destabilization is impossible.

Performance verification: The *exact* LQR design problem for a system with saturating actuator is: Given

$$
\begin{aligned}
\dot{x}_G &= A x_G + B_1 w + B_2 \mathrm{sat}_\alpha(u), \\
z &= C_1 x_G + D_{12} u, \\
u &= K x_G,
\end{aligned}
\tag{6.20}
$$

find K so that

$$
J_{\mathrm{exact}} := \sigma_z^2
\tag{6.21}
$$

is minimized over all K, which render the closed loop system asymptotically stable. As it was pointed out above, the solution of this problem is unknown. Instead, we formulated and solved the following stochastically linearized design problem: Given

$$
\begin{aligned}
\dot{\hat{x}}_G &= (A + B_2 N K)\hat{x}_G + B_1 w, \\
\hat{z} &= (C_1 + D_{12} K)\hat{x}_G, \\
N &= \mathrm{erf}\left(\frac{a}{\sqrt{2}\sigma_{\hat{u}}}\right),
\end{aligned}
\tag{6.22}
$$

find K such that

$$
J_{\mathrm{SLQR}} := \sigma_{\hat{z}}^2
\tag{6.23}
$$

is minimized over all K such that $A + B_2 N K$ is Hurwitz. The solution of this problem is given by K defined in (6.9). We use this controller, denoted below as K_{SLQR}, for the original system (6.3), that is, obtain the following closed loop system:

$$
\begin{aligned}
\dot{\tilde{x}}_G &= A \tilde{x}_G + B_1 w + B_2 \mathrm{sat}_\alpha(\tilde{u}), \\
\tilde{z} &= C_1 \tilde{x}_G + D_{12} \tilde{u}, \\
\tilde{u} &= K_{\mathrm{SLQR}} \tilde{x}_G.
\end{aligned}
\tag{6.24}
$$

The problem of performance verification consists of determining how well $\sigma_{\tilde{z}}^2$ approximates $\min_K J_{\mathrm{exact}}$ and estimating the relative error $|\sigma_{\tilde{z}}^2 - \sigma_{\hat{z}}^2|/\sigma_{\tilde{z}}^2$. These issues can be addressed in both asymptotic and non-asymptotic settings. In the asymptotic case, that is, when disturbance intensity tends to zero, it is possible to prove that $\sigma_{\tilde{z}}^2$ approximates well σ_z^2 and is approximated well by $\sigma_{\hat{z}}^2$. In the non-asymptotic case, it is possible to show by examples that in most cases the relative error is well within 10%. However, in some specifically contrived situations, the relative error may be as large as 20%, but this requires a pathological choice of system parameters.

6.1.4 SLQG Theory

SLQG synthesis equations: Consider the open loop system

$$
\begin{aligned}
\dot{x}_G &= Ax_G + B_1 w + B_2 \text{sat}_\alpha(u), \\
z &= C_1 x_G + D_{12} u, \\
y_m &= C_2 x_G + D_{21} w
\end{aligned}
\tag{6.25}
$$

with the output feedback controller

$$
\begin{aligned}
\dot{x}_C &= M x_C - L y_m, \\
u &= K x_C,
\end{aligned}
\tag{6.26}
$$

where $x_C \in \mathbf{R}^{n_{x_C}}$, and assume the following:

Assumption 6.2.
  (i) (A, B_2) is stabilizable and (C_2, A) is detectable;
  (ii) (A, B_1) is stabilizable and (C_1, A) is detectable;
  (iii) $D_{12} = [0 \ \sqrt{\rho}]^T$, $\rho > 0$ and $D_{21} = [0 \ \sqrt{\mu}]$, $\mu > 0$;
  (iv) $D_{12}^T C_1 = 0$ and $B_1 D_{21}^T = 0$;
  (v) A has no eigenvalues in the open right-half plane.

With the above controller, the closed loop system is

$$
\begin{aligned}
\dot{x}_G &= Ax_G + B_2 \text{sat}_\alpha(Kx_C) + B_1 w, \\
\dot{x}_C &= M x_C - L C_2 x_G - L D_{21} w, \\
z &= C_1 x_G + D_{12} K x_C.
\end{aligned}
\tag{6.27}
$$

Application of stochastic linearization to this system yields

$$
\begin{aligned}
\dot{\hat{x}}_G &= A\hat{x}_G + B_2 N K \hat{x}_C + B_1 w, \\
\dot{\hat{x}}_C &= M\hat{x}_C - L C_2 \hat{x}_G - L D_{21} w, \\
\hat{z} &= C_1 \hat{x}_G + D_{12} K \hat{x}_C, \\
N &= \text{erf}\left(\frac{a}{\sqrt{2}\sigma_{\hat{u}}}\right).
\end{aligned}
\tag{6.28}
$$

Defining

$$
\tilde{A} = \begin{bmatrix} A & B_2 N K \\ -L C_2 & M \end{bmatrix}, \ \tilde{B} = \begin{bmatrix} B_1 \\ -L D_{21} \end{bmatrix},
$$
$$
\tilde{C} = \begin{bmatrix} C_1 & D_{12} K \end{bmatrix}, \ \tilde{K} = \begin{bmatrix} 0 & K \end{bmatrix},
\tag{6.29}
$$

the stochastically linearized state space equations can be rewritten compactly as

$$
\begin{aligned}
\dot{\hat{x}} &= \tilde{A}\hat{x} + \tilde{B}w, \\
\hat{z} &= \tilde{C}\hat{x}, \\
\hat{u} &= \tilde{K}\hat{x},
\end{aligned}
\tag{6.30}
$$

where $\hat{x} = [\hat{x}_G^T \ \hat{x}_C^T]^T$.

Theorem 6.5. *Under Assumption 6.2, there exists a unique controller $C(s) = K(sI - M)^{-1}L$ that solves the SLQG problem*

$$\min_{K,L,M} \sigma_{\tilde{z}}^2, \qquad (6.31)$$

where the minimization is over all (K,L,M) such that $\tilde{A}$ is Hurwitz. Moreover, the minimum value of the cost is

$$\min_{K,L,M} \sigma_{\tilde{z}}^2 = \text{tr}\{C_1(P+R)C_1^T\} + \rho\frac{N^2}{(\rho+\lambda)^2}B_2^T QRQB_2, \qquad (6.32)$$

and a state space realization for the controller is

$$\begin{aligned} K &= -\frac{N}{\rho+\lambda}B_2^T Q, \\ L &= -PC_2^T\frac{1}{\mu}, \\ M &= A + B_2 NK + LC_2, \end{aligned} \qquad (6.33)$$

where (P,Q,R,S,N,λ) is the unique solution of the following system of equations:

$$AP + PA^T - \frac{1}{\mu}PC_2^T C_2 P + B_1 B_1^T = 0, \qquad (6.34)$$

$$A^T Q + QA - \frac{N^2}{\rho+\lambda}QB_2 B_2^T Q + C_1^T C_1 = 0, \qquad (6.35)$$

$$(A - \frac{N^2}{\rho+\lambda}B_2 B_2^T Q)R + R(A - \frac{N^2}{\rho+\lambda}B_2 B_2^T Q)^T + PC_2^T C_2 P\frac{1}{\mu} = 0, \qquad (6.36)$$

$$(A - PC_2^T C_2\frac{1}{\mu})^T S + S(A - PC_2^T C_2\frac{1}{\mu}) + \frac{N^2}{\rho+\lambda}QB_2 B_2^T Q = 0, \qquad (6.37)$$

$$\left(\frac{N^2}{\rho+\lambda}\right)^2 B_2^T QRQB_2 - \frac{N^2 \alpha^2}{2\left[\text{erf}^{-1}(N)\right]^2} = 0, \qquad (6.38)$$

$$\lambda - \frac{\rho}{\frac{\sqrt{\pi}}{2}\frac{N}{\text{erf}^{-1}(N)}\exp\left(\left[\text{erf}^{-1}(N)\right]^2\right) - 1} = 0. \qquad (6.39)$$

In addition, if R and S are nonsingular, then the realization $\{M,L,K\}$ given above is a minimal realization for $C(s)$.

Proof. See Section 8.5.

The proof of this theorem suggests the following bisection algorithm to compute the SLQG controller.

SLQG Bisection Algorithm 6.2 Given a desired accuracy $\epsilon > 0$,

(a) Start with $N_1 = 0$ and $N_2 = 1$
(b) Let $N = (N_1 + N_2)/2$
(c) Calculate λ from (6.39)
(d) Solve the Riccati equations (6.34) and (6.35) for P and Q, respectively
(e) Solve the Lyapunov equations (6.36) and (6.37) for R and S, respectively
(f) Calculate the left-hand side of (6.38) and call it δ
(g) If $|N_2 - N_1| < \epsilon$, go to step (i)
(h) If $\delta < 0$, let $N_1 = N$, else let $N_2 = N$, and go to step (b)
(i) Calculate K, L, M from (6.33).

As in the case of SLQR, this algorithm is convergent and provides the controller $C(s)$ with a desired accuracy.

Similar to the SLQR case, equations (6.32)–(6.39) reduce to the standard LQG equations when the saturation is removed. Note also that since Q depends on P, *the separation principle does not hold for SLQG*.

Structure of the SLQG controller: Below, the structure of the SLQG controller is investigated. Specifically, it is shown that the SLQG controller determined above has the structure of an observer-based controller. In addition, a new *nonlinear* SLQG control law is introduced.

Recall that the controller determined above is given by

$$\dot{x}_C = Mx_C - Ly_m,$$
$$u = Kx_C. \tag{6.40}$$

Substituting M from (6.33), these equations can be written in the form of an observer-based controller:

$$\dot{x}_C = Ax_C + B_2Nu - L(y_m - C_2x_C),$$
$$u = Kx_C. \tag{6.41}$$

This suggest the following *nonlinear* SLQG control law

$$\dot{x}_C = Ax_C + B_2\mathrm{sat}_\alpha(u) - L(y_m - C_2x_C),$$
$$u = Kx_C. \tag{6.42}$$

In this case, although the controller is nonlinear, the estimation error $e = x_G - x_C$ satisfies the usual linear homogeneous equation

$$\dot{e} = (A + LC_2)e, \tag{6.43}$$

and hence, true estimation takes place. Note further that the linear controller (6.41) and the nonlinear controller (6.42) yield exactly the same stochastically linearized performance. Consequently, in the sequel, we use the nonlinear version of the SLQG controller.

Stability verification: Next, we study the stability of the closed loop system formed by the plant (6.25) and the SLQG controller (6.42) in the absence of external input w.

Theorem 6.6. *Consider the undisturbed version of system (6.25):*

$$
\begin{aligned}
\dot{x}_G &= Ax_G + B_2 sat_\alpha(u), \\
y &= C_2 x_G.
\end{aligned}
\tag{6.44}
$$

Let Assumption 6.2 hold and the SLQG controller be given by (6.42). Then

  (i) $(x_G, x_C) = (0,0)$ *is the unique equilibrium point of (6.44), (6.42);*
  (ii) *this equilibrium is exponentially stable;*
  (iii) *if A and M are Hurwitz, all solutions of the closed loop system (6.44), (6.42) are bounded.*

Proof. See Section 8.5.

Performance verification: Performance verification results, similar to those for the SLQR, remain valid for the SLQG case. An example, illustrating the accuracy of stochastic linearization in the SLQG case, is given below.

Example 6.1. Consider the system

$$
\begin{aligned}
\dot{x}_G &= \begin{bmatrix} -1 & -2 & -1 \\ 1 & 0 & 0 \\ 0 & 1 & 0 \end{bmatrix} x_G + \begin{bmatrix} 0 \\ 0 \\ 1 \end{bmatrix} w_1 + \begin{bmatrix} 1 \\ 5 \\ 0 \end{bmatrix} sat(u), \\
z_1 &= \begin{bmatrix} 0 & 1 & 1 \end{bmatrix} x_G, \quad z_2 = \begin{bmatrix} \sqrt{\rho} \end{bmatrix} u, \\
y_m &= \begin{bmatrix} 0 & 0 & 3 \end{bmatrix} x_G + \begin{bmatrix} 1 \times 10^{-4} \end{bmatrix} w_2.
\end{aligned}
\tag{6.45}
$$

The open loop variance of z_1 is 1. Suppose that the specification is to reduce the closed loop variance of z_1 to around 0.025. First, using Theorem 6.1, with a very small ρ, we estimate the smallest achievable $\sigma_{z_1}^2$ as 0.015. Next, we apply Theorem 6.5 to obtain an SLQG controller. With $\rho = 0.0095$, SLQG Solution Algorithm 6.2 results in the controller

$$
C(s) = \frac{616.67s^2 + 589.16s + 589.79}{s^3 + 325.09s^2 + 7.58 \times 10^3 s + 8.66 \times 10^3}.
\tag{6.46}
$$

This controller achieves $\sigma_{z_1}^2 = 0.025$ and $N = 0.745$. Since the plant is filtering, we expect that the variance of z_1 defined by (6.45) and (6.46) is close to 0.025. Simulations show that this variance is, indeed, 0.0254.

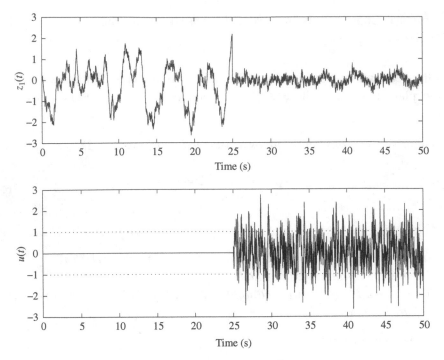

Figure 6.2. Time traces of z_1 and u in system (6.45), (6.46).

When the SLQG controller is implemented as (6.42), the results remain almost the same. For this implementation, the values of K and L are

$$K = \begin{bmatrix} -0.255 & -6.417 & -6.168 \end{bmatrix},$$
$$L = \begin{bmatrix} 0.332 & 0.001 & -100.000 \end{bmatrix}^T. \tag{6.47}$$

Traces of z_1 and u in the closed loop system (6.45), (6.46) are shown in Figure 6.2, where the controller is off for the first 25 seconds and on for the next 25 seconds. The graph of u also includes the clipping level ± 1, indicating that the signal at the input of the plant is the clipped version of u.

6.1.5 Application: Ship Roll Damping Problem

Model: Damping roll oscillation caused by sea waves is of importance for passenger ships. The reason is that roll oscillations lead to passenger discomfort, property damage, and reduced crew effectiveness. There is experimental evidence that passenger discomfort can be measured by the standard deviation of the roll angle. Specifically, the standard deviation of the roll angle should be less than 3 deg for performing intellectual work and less than 2.5 deg for passenger ship comfort.

Several approaches to damping ship roll oscillations have been proposed and implemented. In one of them, the ship is equipped with two actively controlled

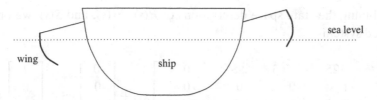

Figure 6.3. Cross section of the ship.

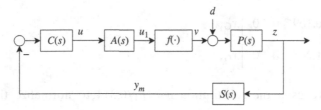

Figure 6.4. System model for ship roll damping problem.

wings attached to its stern (see Figure 6.3). The travel of the wings is limited to a range of ± 18 deg, which is modeled as actuator saturation. We assume that the waves correspond to a sea condition in which the wind velocity is 15.4 m/s, and they are normal to the direction of navigation.

The block diagram of the system, including the ship, actuator, sensor, controller, and wave disturbance, is shown in Figure 6.4. In this figure, $P(s)$ is the ship transfer function from the total torque $v + d$ (N·m) applied by the wings and the wave disturbance to the roll angle z (rad); $A(s)$ is the transfer function from the voltage u (V) applied to the hydraulic actuator that drives the wings to the wings dipping angle u_1 (rad), and $f(u_1)$ is the saturation nonlinearity that relates the wing dipping angle to the torque generated by the wings; $S(s)$ is the transfer function of the gyroscope that is used as a roll angle sensor with output y_m (V); and finally, $C(s)$ is the controller that uses y_m to generate the control voltage u.

From the data provided in the literature, $P(s), A(s), S(s),$ and $f(u_1)$ are as follows:

$$P(s) = \frac{1.484 \times 10^{-6}}{s^2 + 1.125s + 1.563}, \quad A(s) = \frac{0.330}{s + 1.570},$$

$$S(s) = 11.460, \quad f(u_1) = 7.337 \times 10^4 \, \text{sat}\left(\frac{u_1}{0.314}\right). \tag{6.48}$$

We model the disturbance torque generated by the waves as a wide sense stationary (WSS) stochastic process generated by the filter

$$F_1(s) = \frac{7.224 \times 10^4 s}{s^2 + 0.286s + 0.311}, \tag{6.49}$$

driven by standard white noise.

Combining the state space descriptions of $P(s)$, $F_1(s)$, and $S(s)$, we obtain the augmented system

$$
\dot{x}_G = \begin{bmatrix} -1.125 & -1.563 & 0.985 & 0 \\ 1 & 0 & 0 & 0 \\ 0 & 0 & -0.286 & -0.311 \\ 0 & 0 & 1 & 0 \end{bmatrix} x_G + \begin{bmatrix} 0 \\ 0 \\ 1 \\ 0 \end{bmatrix} w_1 + \begin{bmatrix} 1 \\ 0 \\ 0 \\ 0 \end{bmatrix} \text{sat}(u),
$$

(6.50)

$$
z_1 = \begin{bmatrix} 0 & 0.109 & 0 & 0 \end{bmatrix} x_G,
$$

$$
y_1 = \begin{bmatrix} 0 & 1.248 & 0 & 0 \end{bmatrix} x_G,
$$

where we have used the substitution $u = u_1/0.314$ to normalize the saturation characteristic.

The open loop standard deviation of z_1 is 5.55 deg. Since the sea condition considered is very rough, our goal is to reduce the standard deviation of z_1 below 3 deg. Ignoring the saturation, the lead controller

$$
C(s) = 35 \frac{s+0.1}{s+10}
$$

(6.51)

is designed to achieve this goal with $\sigma_{z_1} = 2.64$ deg. However, when this controller is implemented in the actual system, simulations reveal that the standard deviation of z_1 is 3.14 deg, and hence, the performance of this controller does not meet specifications. Of course, reducing the controller gain makes the degradation smaller at the expense of providing poor disturbance rejection. Indeed, the controller

$$
C(s) = 10 \frac{s+0.1}{s+10}
$$

(6.52)

essentially avoids saturation, but achieves an inadequate disturbance rejection since it yields $\sigma_{z_1} = 4.21$ deg.

Design: Using the SLQR Bisection Algorithm 6.1 with a very small ρ, the best achievable level of disturbance rejection, as measured by the standard deviation of the roll angle, is estimated to be 2.08 deg. Recall that our goal is to reduce σ_{z_1} below 3 deg, and here, we aim at obtaining $\sigma_{\hat{z}_1} = 2.5$ deg. Increasing ρ, we find that when $\rho = 4.703 \times 10^{-4}$, $\sigma_{\hat{z}_1} = 2.5$ deg and the corresponding SLQR state feedback gain is

$$
K = \begin{bmatrix} -0.766 & -1.049 & -0.486 & 0.027 \end{bmatrix},
$$

(6.53)

and $N = 0.5375$. Simulations reveal that the value of σ_{z_1} is about 2.54 deg. Thus, this controller achieves the desired performance.

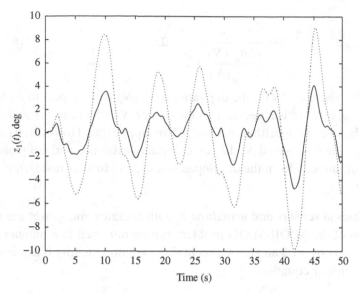

Figure 6.5. Time traces of z_1 with SLQG feedback (—) and open loop ($\cdots$).

Similarly, SLQG Bisection Algorithm 6.2 with $\rho = 1.649 \times 10^{-3}$ and $\mu = 1 \times 10^{-4}$ yields the SLQG controller gains

$$K = \begin{bmatrix} -1.050 & -1.465 & -0.649 & 0.036 \end{bmatrix},$$
$$L = \begin{bmatrix} -28.121 & -6.713 & -87.028 & -36.231 \end{bmatrix}^T. \qquad (6.54)$$

This controller also achieves $\sigma_{\hat{z}_1} = 2.5$ deg. Using simulations, the standard deviation of z_1 is determined as 2.56 deg. A time trace of z_1 with this SLQG controller is shown in Figure 6.5 (the solid curve) together with the corresponding open loop z_1 (the dotted curve). Clearly, the SLQG controller leads to a substantial reduction of ship roll oscillations.

6.1.6 Generalizations

In this subsection, several generalizations of the results obtained above are briefly outlined.

Arbitrary nonlinearities: The methods developed in this section can be readily extended to systems that contain other nonlinearities in the loop, such as deadzone, friction, and quantization. Consider, for example, an arbitrary nonlinearity $f(u)$ in the actuator model. It follows that the equivalent gain N of $f(u)$ is a function of $\sigma_{\hat{u}}^2$, that is, $N = N(\sigma_{\hat{u}}^2)$. Solving for $\sigma_{\hat{u}}^2 = \sigma_{\hat{u}}^2(N)$, it is easy to see that equations (6.12) and (6.13) in Theorem 6.1, and equations (6.38) and (6.39) in Theorem 6.5 take the form

$$\left(\frac{N^2}{\rho + \lambda} \right)^2 B_2^T Q R Q B_2 - N^2 \sigma_{\hat{u}}^2(N) = 0, \qquad (6.12'), (6.38')$$

and

$$\lambda + \frac{\rho}{\dfrac{N}{2}\dfrac{\sigma_{\hat{u}}^{2\prime}(N)}{\sigma_{\hat{u}}^{2}(N)} + 1} = 0, \qquad\qquad (6.13'), (6.39')$$

respectively, where $\sigma_{\hat{u}}^{2\prime}(N)$ is the derivative of $\sigma_{\hat{u}}^{2}(N)$ with respect to N. With these modifications, if $N^2\sigma_{\hat{u}}^2$ is a decreasing function of N, Theorems 6.1 and 6.5 remain valid, and Bisection Algorithms 6.1 and 6.2 are applicable. However, when $N^2\sigma_{\hat{u}}^2$ is not a decreasing function of N, the corresponding existence and uniqueness results do not follow directly from the developed results, and further research is needed to settle them.

Nonlinearities in sensors and actuators: If both actuator and sensor are nonlinear, the solution of the SLQR/SLQG problem can be obtained in a manner similar to that of Subsections 6.1.3 and 6.1.4, except that each nonlinearity is associated with a separate nonlinear equation.

MIMO systems: Throughout the previous subsections, it is assumed that u and y are scalar. When $y \in \mathbf{R}^{n_y}$ is a vector, the extension follows immediately from the previous results by replacing each occurrence of C_2^T/μ with $C_2^T(D_{21}D_{21}^T)^{-1}$, provided that $D_{21}D_{21}^T$ is nonsingular. On the other hand, when $u \in \mathbf{R}^{n_u}$ is a vector, the extension requires multivariable stochastic linearization. It can be shown that the equivalent gain N of a decoupled vector-valued saturation nonlinearity $\mathrm{sat}_A(u) = [\mathrm{sat}_{\alpha_1}(u_1) \cdots \mathrm{sat}_{\alpha_{n_u}}(u_{n_u})]^T$ is given by the diagonal matrix $N = \mathrm{diag}\{N_1,\dots,N_{n_u}\}$, where each N_k is the equivalent gain of the individual nonlinearity $\mathrm{sat}_{\alpha_k}(u_k)$. With this result, the desired extension follows from the previous results by replacing each occurrence of $(N/(\rho+\lambda))B_2^T$ with $(D_{12}^T D_{12} + \Lambda)^{-1}NB_2^T$, where $\Lambda = \mathrm{diag}\{\lambda_1,\dots,\lambda_{n_u}\}$ is the corresponding Lagrange multiplier, provided that $D_{12}^T D_{12}$ is nonsingular and diagonal.

6.2 Instrumented LQR/LQG

6.2.1 Scenario

Section 6.1 provides a method for designing controllers for LPNI systems when the parameters of the instrumentation are given *a priori*. The goal of this section is to develop a method for simultaneous design of controllers and instrumentation. To accomplish this, we parameterize the actuators and sensors by the severity of their nonlinearities, for example, levels of saturation, steps of quantization, and so on. Then, we introduce a performance index, which includes both the system behavior and the parameters of the instrumentation. Assuming that this performance index is quadratic, we derive synthesis equations for designing optimal controllers and instrumentation simultaneously. The resulting technique is referred to as ILQR/ILQG, where the "I" stands for "instrumented".

More specifically, assume, for example, that the actuator and sensor are saturating with the saturation levels α and β, respectively, and consider the system of

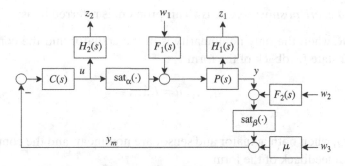

Figure 6.6. LPNI system configuration for ILQR/ILQG problem formulation.

Figure 6.6, where, as usual, $P(s)$ is the plant, $C(s)$ is the controller, $F_1(s)$, and $F_2(s)$ are coloring filters, μ is a static gain used for LQG synthesis, and $H_1(s)$, and $H_2(s)$ are weighting filters. The signals $u, y, y_m \in \mathbf{R}$ are, respectively, the control, plant output, and measured output, while $w_1, w_2, w_3 \in \mathbf{R}$ are independent white noise processes. The controlled (performance) outputs are $z_1, z_2 \in \mathbf{R}$.

The assumption that signals are scalar is made to simplify the technical presentation. This is not restrictive, and generalization to multivariable systems is discussed in Subsection 6.2.4.

Assume that the system, excluding the controller, has the following state space representation:

$$
\begin{aligned}
\dot{x}_G &= A x_G + B_1 w + B_2 \mathrm{sat}_\alpha (u), \\
z &= C_1 x_G + D_{12} u, \\
y &= C_2 x_G, \\
y_m &= \mathrm{sat}_\beta (y) + D_{21} w,
\end{aligned}
\tag{6.55}
$$

where x_G is the state vector, $w = [w_1\ w_2\ w_3]^T$, $z = [z_1\ z_2]^T$, and $D_{21} = [0\ 0\ \mu]$.

The stochastically linearized version of (6.55) is given by

$$
\begin{aligned}
\dot{\hat{x}}_G &= A \hat{x}_G + B_1 w + B_2 N_a \hat{u}, \\
\hat{z} &= C_1 \hat{x}_G + D_{12} \hat{u}, \\
\hat{y}_m &= N_s C_2 \hat{x}_G + D_{21} w,
\end{aligned}
\tag{6.56}
$$

where

$$
\begin{aligned}
N_a &= \mathrm{erf}\left(\frac{\alpha}{\sqrt{2}\sigma_{\hat{u}}}\right), \\
N_s &= \mathrm{erf}\left(\frac{\beta}{\sqrt{2}\sigma_{\hat{y}}}\right).
\end{aligned}
\tag{6.57}
$$

Given (6.56), the design problem considered here is: *Design a controller, and select α and β in order to minimize*

$$
J = \sigma_{\hat{z}}^2 + \eta_a \alpha^2 + \eta_s \beta^2,
$$

where η_a and η_s are positive constants. This problem is referred to as:

- *ILQR*, when the only nonlinearity is in the actuator and the controller is a linear state feedback of the form

$$\hat{u} = K\hat{x}_G; \qquad (6.58)$$

- *ILQG*, when both actuator and sensor are nonlinear, and the controller is an output feedback of the form

$$\dot{\hat{x}}_C = M\hat{x}_C - L\hat{y}_m, \\ \hat{u} = K\hat{x}_C. \qquad (6.59)$$

Since the controllers (6.58) and (6.59), along with instrumentation $\mathrm{sat}_\alpha(\cdot)$ and $\mathrm{sat}_\beta(\cdot)$, are intended to be used in the LPNI system of Figure 6.6, in addition to the above design problems, the following verification problem must be addressed: *With the ILQR/ILQG controller*

$$u = Kx_G \qquad (6.60)$$

or

$$\dot{x}_C = Mx_C - Ly_m, \\ u = Kx_C, \qquad (6.61)$$

and instrumentation $\mathrm{sat}_\alpha(\cdot)$ *and* $\mathrm{sat}_\beta(\cdot)$, *characterize the stability and performance of the resulting closed loop LPNI system.*

The ILQR and ILQG problems, along with their associated verification problems, are considered in Subsections 6.2.2 and 6.2.3, respectively.

6.2.2 ILQR Theory

ILQR synthesis: Consider the open loop LPNI system,

$$\dot{x}_G = Ax_G + B_1 w + B_2\mathrm{sat}_\alpha(u), \\ z = C_1 x_G + D_{12}u, \qquad (6.62)$$

with linear state feedback

$$u = Kx_G. \qquad (6.63)$$

Let Assumption 6.1 of Section 6.1 hold.

From (6.62) and (6.63), the closed loop system is described by

$$\dot{x}_G = Ax_G + B_2\mathrm{sat}_\alpha(Kx_G) + B_1 w, \\ z = (C_1 + D_{12}K)x_G. \qquad (6.64)$$

Applying stochasic linearization to (6.64) results in

$$\dot{\hat{x}}_G = (A + B_2NK)\hat{x}_G + B_1w,$$

$$\hat{z} = (C_1 + D_{12}K)\hat{x}_G,$$

(6.65)

$$N = \text{erf}\left(\frac{\alpha}{\sqrt{2}\sigma_{\hat{u}}}\right).$$

The ILQR Problem is: *Find the value of the gain K and parameter α of the actuator, which ensure*

$$\min_{K,\alpha}\left\{\sigma_{\hat{z}}^2 + \eta\alpha^2\right\}, \eta > 0,$$

(6.66)

where the minimization is over all pairs (K,α) such that $A + B_2NK$ is Hurwitz.

This is a constrained optimization problem, since (6.66) can be rewritten as

$$\min_{K,\alpha}\left\{\text{tr}\left(C_1RC_1^T\right) + \rho KRK^T + \eta\alpha^2\right\},$$

(6.67)

where R satisfies

$$(A + B_2NK)R + R(A + B_2NK)^T + B_1B_1^T = 0,$$

(6.68)

with N defined by

$$KRK^T - \frac{\alpha^2}{2}\left[\text{erf}^{-1}(N)\right]^{-2} = 0.$$

(6.69)

Theorem 6.7. *Under Assumption 6.1, the ILQR problem is solved by*

$$K = -\frac{N}{\lambda + \rho}B_2^TQ,$$

(6.70)

$$\alpha = \text{erf}^{-1}(N)\sqrt{2}\sqrt{KRK^T},$$

(6.71)

where (Q,R,N,λ) is the unique solution of

$$A^TQ + QA - \frac{N^2}{\rho + \lambda}QB_2B_2^TQ + C_1^TC_1 = 0,$$

(6.72)

$$\left(A - \frac{N^2}{\rho + \lambda}B_2B_2^TQ\right)R + R\left(A - \frac{N^2}{\rho + \lambda}B_2B_2^TQ\right)^T + B_1B_1^T = 0,$$

(6.73)

$$\lambda - \frac{\rho}{\frac{N\sqrt{\pi}}{2\text{erf}^{-1}(N)}\exp\left(\text{erf}^{-1}(N)^2\right) - 1} = 0,$$

(6.74)

$$\eta - \frac{\lambda}{2\left(\text{erf}^{-1}(N)\right)^2} = 0,$$

(6.75)

while the optimal ILQR cost is

$$\min_{K,\alpha}\left\{\sigma_{\hat{z}}^2 + \eta\alpha^2\right\} = \mathrm{tr}\left(C_1 R C_1^T\right) + \rho\frac{N^2}{(\rho+\lambda)^2}B_2^T QRQB_2 + 2\eta KRK^T \mathrm{erf}^{-1}(N)^2.$$

$$(6.76)$$

Proof. See Section 8.5.

To find the solution of (6.72)–(6.75), a standard bisection algorithm can be used. Indeed, using (6.74) to substitute for λ in (6.75) yields

$$h(N) - \frac{\rho}{\eta} = 0,\qquad (6.77)$$

where

$$h(N) = N\sqrt{\pi}\,\mathrm{erf}^{-1}(N)\exp\left(\mathrm{erf}^{-1}(N)^2\right) - 2\mathrm{erf}^{-1}(N)^2.\qquad (6.78)$$

It is shown in the proof of Theorem 6.7 that $h(N)$ is continuous and monotonically increasing for $N \in [0,1)$. This leads to the following ILQR solution methodology:

ILQR Bisection Algorithm 6.3: Given a desired accuracy $\epsilon > 0$,

(a) Find an ϵ-precise solution N of (6.77) using bisection (with initial conditions $N^- = 0, N^+ = 1$)
(b) Find λ from (6.74) or (6.75)
(c) Find Q from (6.72)
(d) Find R from (6.73)
(e) Compute K and α from (6.70) and (6.71).

Note that ILQR can be viewed as a generalization of the SLQR theory presented in Section 6.1. Indeed, if α is fixed, (6.75) becomes superfluous and (6.71) becomes a constraint, so that the minimization (6.67) amounts to solution of the SLQR problem.

ILQR performance limitations: Assume that ρ and η are design parameters and denote the first term in the right-hand side of (6.76) as $\gamma^2(\rho,\eta)$, that is,

$$\gamma^2(\rho,\eta) := \mathrm{tr}\left(C_1 R(\rho,\eta) C_1^T\right),\qquad (6.79)$$

where, from (6.73), $R(\rho,\eta) \geq 0$. Note that $\gamma^2(\rho,\eta)$ is the variance of the output $\hat{z}_1$ of the quasilinear system. The following theorem establishes performance limitations on $\gamma^2(\rho,\eta)$.

Theorem 6.8. *Under Assumption 6.1,*

(i) $\gamma^2(\rho,\eta)$ *is an increasing function of ρ and*

$$\lim_{\rho\to 0^+}\gamma^2(\rho,\eta) = \mathrm{tr}\left(C_1 \bar{R}_\eta C_1^T\right),\qquad (6.80)$$

where $(\bar{R}_\eta \geq 0, \bar{Q}_\eta \geq 0)$ *is the unique solution of*

$$\left(A - \frac{2}{\pi\eta}B_2 B_2^T \bar{Q}_\eta\right)\bar{R}_\eta + \bar{R}_\eta\left(A - \frac{2}{\pi\eta}B_2 B_2^T \bar{Q}_\eta\right)^T + B_1 B_1^T = 0, \quad (6.81)$$

$$A^T \bar{Q}_\eta + \bar{Q}_\eta A - \frac{2}{\pi\eta}\bar{Q}_\eta B_2 B_2^T \bar{Q}_\eta + C_1^T C_1 = 0; \quad (6.82)$$

(ii) $\gamma^2(\rho,\eta)$ *is an increasing function of* η *and*

$$\lim_{\eta \to 0^+} \gamma^2(\rho,\eta) = \gamma_{\rho 0}^2, \quad (6.83)$$

where $\gamma_{\rho 0}^2$ *denotes the optimal output variance achievable by conventional LQR, that is, with linear instrumentation;*

(iii) *if* A *is Hurwitz,*

$$\lim_{\rho \to \infty} \gamma^2(\rho,\eta) = \lim_{\eta \to \infty} \gamma^2(\rho,\eta) = \gamma_{OL}^2, \quad (6.84)$$

where γ_{OL}^2 *denotes the open loop output variance of* z_1.

Proof. See Section 8.5.

Thus, for any η, the output variance cannot be made smaller than $\operatorname{tr}\left\{C_1 \bar{R}_\eta C_1^T\right\}$; for a given ρ, the output variance cannot be made smaller than $\gamma_{\rho 0}^2$; as η and ρ tend to ∞, the optimal strategy is to operate the system in an open loop regime.

ILQR stability verification: The problem of ILQR stability verification consists of investigating the stability properties of the closed loop LPNI system (6.64), with ILQR controller (6.70) and instrumentation (6.71). To address this problem, consider the following undisturbed version of (6.64):

$$\begin{aligned}
\dot{x}_G &= A x_G + B_2 \mathrm{sat}_\alpha(K x_G), \\
z &= (C_1 + D_{12}K)x_G.
\end{aligned} \quad (6.85)$$

Assume that the pair (K,α) is obtained from (6.70) and (6.71), and (Q,R,N,λ) is the corresponding solution of (6.72)–(6.75).

Theorem 6.9. *For the closed loop system (6.85) with (6.70), (6.71):*

(i) $x_G = 0$ *is the unique equilibrium;*
(ii) *this equilibrium is exponentially stable;*
(iii) *a subset of its domain of attraction is given by*

$$\mathcal{X} = \left\{x_G \in R^{n_x} : x_G^T\left(QB_2 B_2^T Q\right)x_G \leq \frac{4(\rho+\lambda)^2 \alpha^2}{N^4}\right\}. \quad (6.86)$$

Proof. See Section 8.5.

The additive disturbance w may force the system to exit the domain of attraction. One may use tools from absolute stability theory, such as the Popov stability criterion, to verify that destabilization by such a disturbance is impossible.

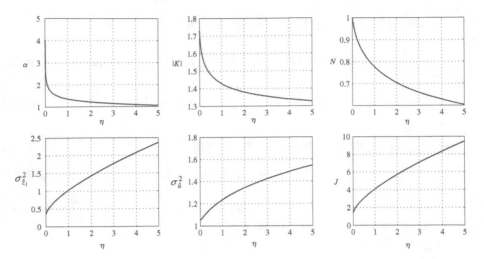

Figure 6.7. ILQR solution for a double integrator as a function of actuator penalty η.

Illustrative example: The following example illustrates the behavior of the ILQR solution as the instrumentation penalty η is increased. Consider the standard double-integrator plant in the form (6.62), with system matrices

$$A = \begin{bmatrix} 0 & 0 \\ 1 & 0 \end{bmatrix}, B_1 = B_2 = \begin{bmatrix} 1 \\ 0 \end{bmatrix}$$

$$C_1 = \begin{bmatrix} 0 & 1 \\ 0 & 0 \end{bmatrix}, D_{12} = \begin{bmatrix} 0 \\ \sqrt{\rho} \end{bmatrix}, \tag{6.87}$$

and assume that $\rho = 1$.

Figure 6.7 shows the behavior of the ILQR solution for a range of η. Clearly, as η approaches 0^+, α and $|K|$ increase, the equivalent gain N tends to 1, and the ILQR solution coincides with the conventional LQR solution. As η increases, the output variances $\sigma_{\hat{z}_1}^2$ and $\sigma_{\hat{u}}^2$ increase, as does the optimal ILQR cost J. This is consistent with Theorem 6.8, noting that the double-integrator plant (6.87) does not have a finite open loop steady state variance.

6.2.3 ILQG Theory

ILQG synthesis equations: Consider the open loop LPNI system of Figure 6.6 represented as

$$\begin{aligned}
\dot{x}_G &= Ax_G + B_1 w + B_2 \text{sat}_\alpha (u), \\
z &= C_1 x_G + D_{12} u, \\
y &= C_2 x_G, \\
y_m &= \text{sat}_\beta (y) + D_{21} w,
\end{aligned} \tag{6.88}$$

with output feedback controller

$$\dot{x}_C = Mx_C - Ly_m,$$
$$u = Kx_C. \tag{6.89}$$

Let Assumption 6.2 of Section 6.1 hold. Furthermore, assume that $F_2(s)$ is strictly proper. This ensures that $D_{21}w$ is the only direct noise feedthrough onto the measured output y_m.

From (6.88) and (6.89), the closed loop LPNI system is

$$\dot{x}_G = Ax_G + B_1w + B_2\text{sat}_\alpha(Kx_C),$$
$$\dot{x}_C = Mx_C - L\left(\text{sat}_\beta(C_2x_G) + D_{21}w\right), \tag{6.90}$$
$$z = C_1x_G + D_{12}Kx_C.$$

Applying stochastic linearization to (6.90) results in

$$\dot{\hat{x}}_G = A\hat{x}_G + B_1w + B_2N_aK\hat{x}_C,$$

$$\dot{\hat{x}}_C = M\hat{x}_C - LN_sC_2\hat{x}_G - LD_{21}w,$$

$$\hat{z} = C_1\hat{x}_G + D_{12}K\hat{x}_C,$$

$$\hat{y}_m = N_sC_2\hat{x}_G + D_{21}w, \tag{6.91}$$

$$N_a = \text{erf}\left(\frac{\alpha}{\sqrt{2}\sigma_{\hat{u}}}\right),$$

$$N_s = \text{erf}\left(\frac{\beta}{\sqrt{2}\sigma_{\hat{y}}}\right),$$

which can be rewritten as

$$\dot{\hat{x}} = \left(\tilde{A} + \tilde{B}_2\tilde{N}\tilde{C}_2\right)\hat{x} + \tilde{B}_1w,$$
$$\hat{z} = \tilde{C}_1\hat{x}, \tag{6.92}$$

where

$$\tilde{A} = \begin{bmatrix} A & 0 \\ 0 & M \end{bmatrix}, \tilde{N} = \begin{bmatrix} N_a & 0 \\ 0 & N_s \end{bmatrix},$$

$$\tilde{B}_1 = \begin{bmatrix} B_1 \\ -LD_{21} \end{bmatrix}, \tilde{C}_1 = \begin{bmatrix} C_1 & D_{12}K \end{bmatrix}, \tag{6.93}$$

$$\tilde{B}_2 = \begin{bmatrix} B_2 & 0 \\ 0 & -L \end{bmatrix}, \tilde{C}_2 = \begin{bmatrix} 0 & K \\ C_2 & 0 \end{bmatrix},$$

and $\hat{x} = [\hat{x}_G^T \ \hat{x}_C^T]^T$.

The ILQG Problem is: *Find the values of $K, L, M, \alpha,$ and β, which ensure*

$$\min_{K,L,M,\alpha,\beta} \left\{\sigma_{\hat{z}}^2 + \eta_a\alpha^2 + \eta_s\beta^2\right\}, \eta_a > 0, \eta_s > 0, \tag{6.94}$$

where the minimization is over all (K,L,M,α,β) such that $(\tilde{A} + \tilde{B}_2\tilde{N}\tilde{C}_2)$ is Hurwitz.

Similar to the ILQR case, this problem can be rewritten as

$$\min_{K,L,M,\alpha,\beta} \left\{ \operatorname{tr}\left(\tilde{C}_1 \tilde{P} \tilde{C}_1^T\right) + \eta_a \alpha^2 + \eta_s \beta^2 \right\}, \tag{6.95}$$

where $\tilde{P}$ satisfies

$$\left(\tilde{A} + \tilde{B}_2 \tilde{N} \tilde{C}_2\right)\tilde{P} + \tilde{P}\left(\tilde{A} + \tilde{B}_2 \tilde{N} \tilde{C}_2\right)^T + \tilde{B}_1 \tilde{B}_1^T = 0 \tag{6.96}$$

with $\tilde{N}$ defined by

$$\operatorname{diag}\left\{\tilde{C}_2 \tilde{P} \tilde{C}_2^T\right\} - \frac{1}{2}\Theta\left[\operatorname{erf}^{-1}\left(\tilde{N}\right)\right]^{-2} = 0, \tag{6.97}$$

$$\Theta = \begin{bmatrix} \alpha^2 & 0 \\ 0 & \beta^2 \end{bmatrix}. \tag{6.98}$$

Theorem 6.10. *Under Assumption 6.2, the ILQG problem (6.94) is solved by*

$$K = -\frac{N_a}{\lambda_1 + \rho} B_2^T Q, \tag{6.99}$$

$$L = -PC_2^T \frac{N_s}{\mu}, \tag{6.100}$$

$$M = A + B_2 N_a K + L N_s C_2, \tag{6.101}$$

$$\alpha = \operatorname{erf}^{-1}(N_a)\sqrt{2}\sqrt{KRK^T}, \tag{6.102}$$

$$\beta = \operatorname{erf}^{-1}(N_s)\sqrt{2}\sqrt{C_2(P+R)C_2^T}, \tag{6.103}$$

where $(P,Q,R,S,N_a,N_s,\lambda_1,\lambda_2)$ *is a solution of*

$$AP + PA^T - \left(\frac{N_s^2}{\mu}\right)PC_2^T C_2 P + B_1 B_1^T = 0, \tag{6.104}$$

$$A^T Q + QA - \left(\frac{N_a^2}{\rho + \lambda_1}\right)QB_2 B_2^T Q + C_1^T C_1 + \lambda_2 C_2^T C_2 = 0, \tag{6.105}$$

$$(A + B_2 N_a K)R + R(A + B_2 N_a K)^T + \mu LL^T = 0, \tag{6.106}$$

$$(A + L N_s C_2)^T S + S(A + L N_s C_2) + \rho K^T K = 0, \tag{6.107}$$

$$\lambda_1 - \frac{\rho}{\frac{N_a \sqrt{\pi}}{2 \operatorname{erf}^{-1}(N_a)} \exp\left(\operatorname{erf}^{-1}(N_a)^2\right) - 1} = 0, \tag{6.108}$$

$$\left(C_2 P S P C_2^T\right) N_s^T \mu - \frac{\sqrt{\pi} \lambda_2 \beta^2}{4} \operatorname{erf}^{-1}(N_s)^{-3} \times \exp\left(\operatorname{erf}^{-1}(N_s)^2\right) = 0, \tag{6.109}$$

$$\eta_a - \frac{\lambda_1}{2\left(\operatorname{erf}^{-1}(N_a)\right)^2} = 0, \tag{6.110}$$

$$\eta_s - \frac{\lambda_2}{2\left(\operatorname{erf}^{-1}(N_s)\right)^2} = 0, \tag{6.111}$$

which minimizes the ILQG cost

$$J_{ILQG} = \operatorname{tr}\left(C_1(P+R)C_1^T\right) + \rho \frac{N^2}{(\rho+\lambda)^2} B_2^T Q R Q B_2 + 2\eta_a K R K^T \operatorname{erf}^{-1}(N_a)^2$$

$$+ 2\eta_s C_2(P+R)C_2^T \operatorname{erf}^{-1}(N_s)^2. \tag{6.112}$$

Proof. See Section 8.5.

The following technique can be used to obtain the ILQG solution:

ILQG Bisection Algorithm 6.4: Given a desired accuracy $\epsilon > 0$,

(a) With $h(\cdot)$ defined in (6.78), find an ϵ-precise solution N_a of the equation

$$h(N_a) - \frac{\rho}{\eta_a} = 0, \tag{6.113}$$

using bisection (with initial conditions $N_a^- = 0, N_a^+ = 1$)

(b) Find λ_1 from (6.108) or (6.110)

(c) For any N_s, the left-hand side of (6.109) can now be determined by finding λ_2, P, Q, R, and S, by solving, in sequence, (6.111), (6.104), (6.105), (6.106), and (6.107). Hence, the left-hand side of (6.109) can be expressed as function of N_s

(d) Find all $N_s \in [0,1]$ that satisfy (6.109) by using a root-finding technique such as numerical continuation or generalized bisection

(e) For each N_s found in the previous step, compute K, L, M, α, β from (6.99)–(6.103)

(f) Find the quintuple (K, L, M, α, β) that minimizes J_{ILQG} from (6.112).

Note that the optimal ILQG solution may correspond to operating openloop, that is, $\alpha, \beta = 0$ and $K(sI - A)^{-1}L \equiv 0$.

Also, in contrast to conventional LQG, due to the interdependence of (6.104)–(6.111) on both N_a and N_s, *the separation principle does not hold for ILQG*.

ILQG controller structure: The ILQG controller (6.89) can be rewritten in the standard observer-based form as

$$\dot{x}_C = Ax_C + B_2 N_a u - L(y_m - N_s C_2 x_C),$$
$$u = Kx_C,$$

(6.114)

which suggests the following nonlinear implementation:

$$\dot{x}_C = Ax_C + B_2 \text{sat}_\alpha(u) - L(y_m - \text{sat}_\beta(C_2 x_C)),$$
$$u = Kx_C.$$

(6.115)

With (6.115), in the absence of the disturbance w, the estimation error $e = x_G - x_C$ satisfies

$$\dot{e} = Ae + L(\text{sat}_\beta(C_2 x_G) - \text{sat}_\beta(C_2 x_C)),$$

(6.116)

so that when $C_2 x_C$ and $C_2 x_G$ are sufficiently small,

$$\dot{e} = (A + LC_2)e,$$

(6.117)

and true estimation occurs. Note that (6.114) and (6.115) yield the same quasilinear closed loop performance.

ILQG stability verification: We now verify the stability of the LPNI system (6.88) controlled by the nonlinear ILQG controller (6.115) with ILQG instrumentation (6.102)–(6.103). Consider the undisturbed version of (6.88) with the controller (6.115), that is,

$$\dot{x}_G = Ax_G + B_2 \text{sat}_\alpha(Kx_C),$$
$$\dot{x}_C = Ax_C + B_2 \text{sat}_\alpha(Kx_C) - L(y_m - \text{sat}_\beta(C_2 x_C)),$$
$$y_m = \text{sat}_\beta(C_2 x_G),$$
$$z = C_1 x_G + D_{12} K x_C.$$

(6.118)

Assume that (K, L, M, α, β) is the ILQG solution (6.99)–(6.103), and

$$(P, Q, R, S, N_a, N_s, \lambda_1, \lambda_2)$$

is the corresponding solution of (6.104)–(6.111). We have the following:

Theorem 6.11. *For the closed loop system (6.118) with (6.99)–(6.103):*

(i) $[x_G, x_C] = 0$ is the unique equilibrium;

(ii) this equilibrium is exponentially stable;

(iii) if $x_G(0) = x_C(0)$, a subset of its domain of attraction is given by the set $\mathcal{Y} \times \mathcal{Y}$, where

$$\mathcal{Y} = \left\{ x_G \in R^{n_x} : x_G^T \left(QB_2 B_2^T Q \right) x_G \le \frac{4(\rho + \lambda_1)^2 \alpha^2}{N_a^4} \right\} \cap$$
$$\left\{ x_G \in R^{n_x} : x_G^T C_2^T C_2 x_G \le \beta^2 \right\}.$$

(6.119)

Proof. See Section 8.5.

Illustrative example: Reconsider the double integrator example of Subsection 6.2.2, represented in the form (6.88), with A, B_2, C_1, D_{12} as defined in (6.87) and

$$B_1 = \begin{bmatrix} 0 & 0 \\ 1 & 0 \end{bmatrix}, D_{21} = \begin{bmatrix} 0 & \sqrt{\mu} \end{bmatrix}, C_2 = C_1. \qquad (6.120)$$

Assume that $\rho = \mu = 1 \times 10^{-4}$ and consider the following two cases:

Case 1: Fix $\eta_s = 2 \times 10^{-3}$ and examine $\sigma_{\hat{z}_1}^2, \alpha, \beta$ as functions of η_a.
Case 2: Fix $\eta_a = 2 \times 10^{-3}$ and examine $\sigma_{\hat{z}_1}^2, \alpha, \beta$ as functions of η_s.

Figure 6.8 illustrates the ILQG solution for Case 1, where the η_a axis is displayed on a logarithmic scale. Clearly, as η_a increases, α decreases and $\sigma_{\hat{z}_1}^2$ increases. Note that β also decreases, showing that the synthesis of actuator and sensor are not decoupled.

Figure 6.9 shows the behavior for Case 2. As expected, when η_s increases, β decreases and $\sigma_{\hat{z}_1}^2$ increases. In addition, observe that α increases, again demonstrating the lack of separation in ILQG.

6.2.4 Generalizations

Arbitrary nonlinearities: Similar to SLQR/SLQG, the ILQR/ILQG methodologies can be extended to nonlinearities other than saturation. For such nonlinearities, the Lagrange multiplier technique is applied as before, but using different expressions

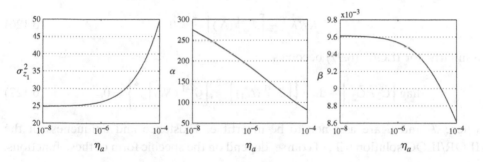

Figure 6.8. ILQG solution for a double integrator as a function of actuator penalty η_a.

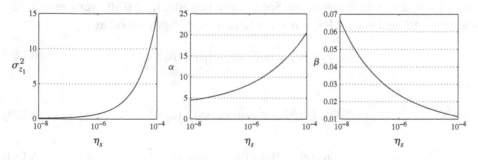

Figure 6.9. ILQG solution for a double integrator as a function of sensor penalty η_s.

for the quasilinear gains. If $f_\alpha(u)$ and $g_\beta(y)$ are the actuator and sensor nonlinearities, the general expressions for the quasilinear gains are

$$N_a = \mathcal{F}(\sigma_{\hat{u}}) = \int_{-\infty}^{\infty} f_\alpha'(x) \frac{1}{\sqrt{2\pi}\,\sigma_{\hat{u}}} \exp\left(-\frac{x^2}{2\sigma_{\hat{u}}^2}\right) dx, \tag{6.121}$$

$$N_s = \mathcal{G}(\sigma_{\hat{y}}) = \int_{-\infty}^{\infty} g_\beta'(x) \frac{1}{\sqrt{2\pi}\,\sigma_{\hat{y}}} \exp\left(-\frac{x^2}{2\sigma_{\hat{y}}^2}\right) dx. \tag{6.122}$$

For example, if $g_\beta(y)$ is the symmetric deadzone defined in (2.23),

$$\mathcal{G}(\sigma_{\hat{y}}) = 1 - \mathrm{erf}\left(\frac{\beta/2}{\sqrt{2}\sigma_{\hat{y}}}\right). \tag{6.123}$$

Similarly, if $g_\beta(y)$ is the quantizer (2.35) then

$$\mathcal{G}(\sigma_{\hat{y}}) = Q_m\left(\frac{\beta}{\sqrt{2}\sigma_{\hat{y}}}\right), \tag{6.124}$$

where

$$Q_m(z) := \frac{2z}{\sqrt{\pi}}\left[\sum_{k=1}^{m} e^{-\frac{1}{4}(2k-1)^2(z)^2}\right]. \tag{6.125}$$

For the ILQR problem (6.67), the constraint (6.69) now becomes

$$KRK^T - \left[\mathcal{F}^{-1}(N)\right]^{-2} = 0. \tag{6.126}$$

Similarly, for ILQG, (6.97) becomes

$$\mathrm{diag}\left\{\tilde{C}_2\check{P}\tilde{C}_2^T\right\} - \mathrm{diag}\left\{\left[\mathcal{F}^{-1}(N_a)\right]^{-2}, \left[\mathcal{G}^{-1}(N_s)\right]^{-2}\right\} = 0, \tag{6.127}$$

where $\mathcal{F}$ and $\mathcal{G}$ are assumed to be invertible. Existence and uniqueness of the ILQR/ILQG solution will, of course, depend on the specific form of these functions.

MIMO systems: The methods presented in Subsections 6.2.2 and 6.2.3 can be extended to the multivariable case. Specifically, consider the MIMO version of (6.88), where $u \in \mathbf{R}^p$ and $y, y_m \in \mathbf{R}^q$, $p, q > 1$, where α, β are understood as

$$\alpha := [\alpha_1 \dots \alpha_p]^T, \ \beta := [\beta_1 \dots \beta_q]^T, \tag{6.128}$$

and

$$\mathrm{sat}_\alpha(u) := [\mathrm{sat}_{\alpha_1}(u_1) \dots \mathrm{sat}_{\alpha_p}(u_p)]^T, \tag{6.129}$$

$$\mathrm{sat}_\beta(y) := [\mathrm{sat}_{\beta_1}(y_1) \dots \mathrm{sat}_{\beta_q}(y_q)]^T. \tag{6.130}$$

As before, the quasilinearization of this system is given by (6.91)–(6.93) with the equivalent gains of (6.129) and (6.130) specified by

$$N_a = \text{diag}\left(N_{a_1}, N_{a_2}, ..., N_{a_p}\right),$$

(6.131)

and

$$N_s = \text{diag}\left(N_{s_1}, N_{s_2}, ..., N_{s_q}\right),$$

(6.132)

respectively, where

$$N_{a_k} = \text{erf}\left(\frac{\alpha_k}{\sqrt{2}\sigma_{\hat{u}_k}}\right), \quad N_{s_l} = \text{erf}\left(\frac{\beta_l}{\sqrt{2}\sigma_{\hat{y}_l}}\right),$$

(6.133)

for $k = 1, ..., p$ and $l = 1, ..., q$.

The ILQG problem (6.94) now becomes

$$\min_{K,L,M,\alpha,\beta} \left\{\sigma_{\tilde{z}}^2 + \alpha^T W_a \alpha + \beta^T W_s \beta\right\},$$

(6.134)

where W_a, W_s are diagonal and positive definite. Clearly, this can be rewritten as

$$\min_{K,L,M,\alpha,\beta} \left\{\text{tr}\left(\tilde{C}_1 \tilde{P} \tilde{C}_1^T\right) + \alpha^T W_a \alpha + \beta^T W_s \beta\right\},$$

(6.135)

subject to the constraints (6.96) and (6.97), with Θ in (6.97) becoming

$$\Theta = \begin{bmatrix} \text{diag}(\alpha\alpha^T) & 0 \\ 0 & \text{diag}(\beta\beta^T) \end{bmatrix}.$$

(6.136)

The optimization is carried out in a manner analagous to the proof of Theorem 6.10, and the necessary conditions for minimality are obtained in terms of the Lagrange multiplier $\Lambda = [\lambda_1, ..., \lambda_{(p+q)}]$.

6.2.5 Application: Ship Roll Damping Problem

Model and problem: The model here remains the same as in Subsection 6.1.5, but with a saturating sensor (along with saturating actuator, see Figure 6.10), that is,

$$\dot{x}_G = \begin{bmatrix} -1.125 & 1.563 & 0.985 & 0 \\ 1 & 0 & 0 & 0 \\ 0 & 0 & -0.286 & -0.311 \\ 0 & 0 & 1 & 0 \end{bmatrix} x_G + \begin{bmatrix} 0 \\ 0 \\ 1 \\ 0 \end{bmatrix} w_1 + \begin{bmatrix} 1 \\ 0 \\ 0 \\ 0 \end{bmatrix} \text{sat}_\alpha(u),$$

(6.137)

$$z_1 = \begin{bmatrix} 0 & 0.109 & 0 & 0 \end{bmatrix} x_G,$$

(6.138)

$$y = \begin{bmatrix} 0 & 1.248 & 0 & 0 \end{bmatrix} x_G,$$

(6.139)

$$y_m = \text{sat}_\beta(y) + \sqrt{\mu} w_2.$$

(6.140)

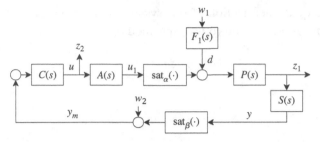

Figure 6.10. Block diagram for ship roll example.

Note that the system is normalized so that $\alpha = 1$ corresponds to an angular travel of 18 degrees. Below, we demonstrate the utility of ILQR/ILQG as compared with SLQR/SLQG.

ILQR solution: The following design objectives are specified:

(1) $\sigma_{\hat{z}_1} < 3$ rad
(2) $\alpha \leq 1$.

Using the ILQR solution method with the penalties $\eta = 3.5 \times 10^{-3}$ and $\rho = 1 \times 10^{-6}$, we obtain

$$K = \begin{bmatrix} -5.641 & -7.565 & -3.672 & 0.2058 \end{bmatrix}, \tag{6.141}$$

$$\alpha = 0.78 \Rightarrow 14 \text{ deg}, \tag{6.142}$$

resulting in

$$\sigma_{\hat{z}_1} = 2.72. \tag{6.143}$$

Numerical simultation of the nonlinear system with this controller and actuator reveals that

$$\sigma_{z_1} = 2.79, \tag{6.144}$$

which verifies the accuracy of the quasilinearization. Clearly, the design objectives are met. Note that by simultaneously synthesizing the controller and instrumentation, we find a solution that uses a saturation authority of less than 18 degrees.

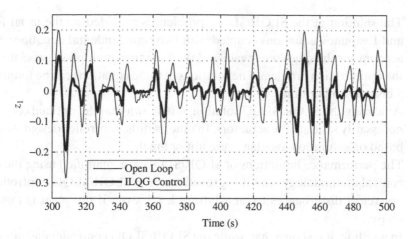

Figure 6.11. Time trace of z_1 for ship-roll example.

ILQG solution: Using the parameters $\eta_a = 2.55 \times 10^{-3}$, $\eta_s = 1 \times 10^{-10}$, $\rho = 1 \times 10^{-5}$, and $\mu = 1 \times 10^{-5}$, the ILQG solution method results in

$$K = \begin{bmatrix} -2.029 & -2.798 & -1.264 & 0.0709 \end{bmatrix}, \tag{6.145}$$

$$L = \begin{bmatrix} -80.77 & -16.09 & -281.41 & -100.38 \end{bmatrix}^T, \tag{6.146}$$

$$\alpha = 0.91 \Rightarrow 16.4\text{deg}, \quad \beta = 0.35 \Rightarrow 6.3\text{deg}, \tag{6.147}$$

leading to

$$\sigma_{\hat{z}_1} = 2.56. \tag{6.148}$$

Simulation of the nonlinear system yields

$$\sigma_{z_1} = 2.77, \tag{6.149}$$

which meets the performance specification. As anticipated, the performance of the actual nonlinear system (6.149) is within 8% of its quasilinear approximation (6.148). Figure 6.11 shows a time trace of z_1 for both the open and closed loop systems. Note that in open loop $\sigma_{z_1} = 5.5$, so that control results in a 50% performance improvement.

6.3 Summary

- The LQR/LQG design methodology can be extended to LPNI systems. This is accomplished by applying the LQR/LQG approach to stochastically linearized versions of the systems at hand. For the case of saturating actuators, the resulting technique is referred to as SLQR/SLQG.

- The solution of the SLQR/SLQG problem is provided by the usual Ricatti and Lyapunov equations coupled with two transcendental equations, which account for the quasilinear gain and the Lagrange multiplier associated with the optimization problem. The solutions of these equations can be found using a bisection algorithm.
- Analyzing the SLQR/SLQG solution, it is shown that the optimal controllers necessarily saturate the actuators. In other words, controller design strategies based on saturation avoidance are not optimal.
- The performance limitations of SLQR/SLQG are quantified using the cheap control methodology, and it is proved that the SLQR/SLQG controllers do not reject disturbances to an arbitrary level, even if the plant is minimum phase.
- In addition, it is shown that, while the SLQR/SLQG controller ensures global asymptotic stability of the stochastically linearized system, the actual, LPNI system inherits this property in a weaker, semi-global stability sense.
- A new problem of state space design for LPNI systems is the so-called ILQR/ILQG problem, where the "I" stands for "Instrumented." Here, the usual LQR/LQG cost is augmented by additional terms representing the instrumentation parameters. As a result, the ILQR/ILQG methodology allows for designing the optimal controller and instrumentation simultaneously.
- The solution of the ILQR/ILQG problem is given by coupled Ricatti, Lyapunov, and transcendental equations. A bisection algorithm to calculate these solutions with any desired accuracy is provided.
- The extent to which the actual, LPNI systems with ILQR/ILQG controllers inherit the stability properties and performance of the stochastically linearized systems is similar to those of SLQR/SLQG.
- The SLQR/SLQG method provides a state space solution of the Narrow Sense Design problem introduced in Section 1.2.
- The ILQR/ILQG method provides a state space solution of the Wide Sense Design problem introduced in Section 1.2.

6.4 Problems

Problem 6.1. An automotive active suspension system is shown in Figure 6.12. Here u is the suspension force provided by a hydraulic actuator. The disturbance d is the road profile. Such a system can be modeled as

$$
\begin{bmatrix} \dot{x}_1 \\ \dot{x}_2 \\ \dot{v}_1 \\ \dot{v}_2 \end{bmatrix} = \begin{bmatrix} 0 & 0 & 1 & 0 \\ 0 & 0 & 0 & 1 \\ -10 & -10 & -2 & 2 \\ 60 & -660 & 12 & -12 \end{bmatrix} \begin{bmatrix} x_1 \\ x_2 \\ v_1 \\ v_2 \end{bmatrix} + \begin{bmatrix} 0 \\ 0 \\ 0.00334 \\ -0.02 \end{bmatrix} u + \begin{bmatrix} 0 \\ 0 \\ 0 \\ 600 \end{bmatrix} d.
$$

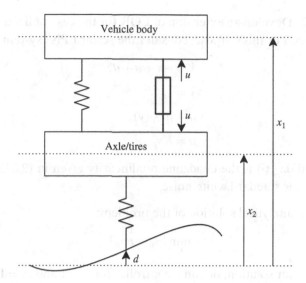

Figure 6.12. Problem 6.1.

For simplicity, assume that d is standard white noise. The control objective is to ensure good ride quality, in terms of the standard deviation of x_1, by rejecting the disturbance.

(a) Use conventional LQR to design a state feedback controller in order to achieve 90% disturbance rejection.

(b) Assume that the actuator can deliver a maximum force of $\pm 10N$. Use stochastic linearization to determine the disturbance rejection performance degradation of the design obtained in (a).

(c) Now, use SLQR to design a state feedback controller to achieve 90% disturbance rejection in the presence of the above saturating actuator.

(d) Find an estimate for the region of attraction of your design.

(e) Suppose that the outputs of the system are defined as $y_1 = x_1 - x_2$ and $y_2 = v_1 - v_2$. Use SLQG to design an output feedback controller to recover the performance obtained in part (c).

(f) Simulate the SLQG design and plot the output response.

(g) A more realistic model for the disturbance is given by colored noise of bandwidth $\Omega = V$, where V is the velocity of the car.

   i. Based on this Ω, model the disturbance as the output of the filter

$$F_\Omega(s) = \frac{\Omega}{s + \Omega} \tag{6.150}$$

   driven by white noise. Redesign your SLQG controller.

   ii. Given this disturbance model, is it possible to use an actuator with a lower maximum force?

Problem 6.2. Develop an extension of LQR for the case of linear actuators and nonlinear sensors. For this purpose, consider the scalar LPNI system specified by

$$\dot{x} = -x + u + d,$$

$$y = x,$$

$$y_m = dz_\Delta(y),$$

$$u = Ky_m,$$

where $x \in \mathbf{R}$ and $dz_\Delta(\cdot)$ is the deadzone nonlinearity given in (2.23). Assume that the disturbance d is standard white noise.

(a) Find an analytical solution of the problem:

$$\min_K (\sigma_y + \rho\sigma_u), \rho > 0.$$

(b) Using your solution, obtain a controller for $\rho = 1$ and $\Delta = 1$.
(c) Simulate the LPNI system with your controller design. Does the simulated performance match the intended design?

Problem 6.3. Develop an extension of SLQR for nonlinearities in both actuators and sensors. Specifically, consider the scalar LPNI system given by

$$\dot{x} = -x + sat_\alpha(u) + d,$$

$$y = x,$$

$$y_m = dz_\Delta(y),$$

$$u = Ky_m,$$

where $x \in \mathbf{R}$ and $dz_\Delta(\cdot)$ is the deadzone nonlinearity given in (2.23). Find an analytical solution of the problem:

$$\min_K (\sigma_y^2 + \rho\sigma_u^2), \rho > 0.$$

Problem 6.4. Consider the active suspension of Problem 6.1 where the disturbance d is standard white noise. The hydraulic actuator is to be chosen to minimize its cost while maintaining an acceptable level of disturbance rejection.

(a) Use ILQR to design a state feedback controller and an actuator that achieves 85% disturbance rejection.
(b) By varying the ILQR penalties, plot tradeoff curves that illustrate the disturbance rejection performance versus the maximum actuator force. What is the smallest saturation level required to achieve 99% disturbance rejection?

Problem 6.5. A servo model is given by

$$
\begin{bmatrix} \dot{\theta} \\ \dot{\omega} \\ \dot{i} \end{bmatrix} =
\begin{bmatrix} 0 & 1 & 0 \\ 0 & 0 & 4.438 \\ 0 & -12 & -24 \end{bmatrix}
\begin{bmatrix} \theta \\ \omega \\ i \end{bmatrix} +
\begin{bmatrix} 0 \\ 0 \\ 20 \end{bmatrix} v +
\begin{bmatrix} 0 \\ -7.396 \\ 0 \end{bmatrix} T_L
$$

where θ is the servo angle, ω is the angular velocity, and i is the current. The input to the servo is the voltage v, while T_L is the load torque. The control objective is to reject random fluctuations in T_L.

(a) Assume that T_L is standard white noise. Use ILQR to choose a voltage saturation level and find a state feedback controller that achieves 90% disturbance rejection.

(b) Suppose that the only measured output is the servo angle, that is, $y = \theta$. Use ILQG to choose a voltage saturation level and find an output feedback controller that achieves 90% disturbance rejection. How does the voltage saturation level compare with that from part (a).

(c) Now, assume that, in addition to the actuator, the measured signal is saturated, that is, $y_m = \text{sat}_\beta(y)$. Use ILQG to design an output feedback controller that achieves 85% disturbance rejection. What are the actuator and sensor specifications, that is, values of α and β.

(d) Simulate the system to evaluate the performance of your controller with respect to other disturbances: sinusoidal, square-wave, sawtooth. Plot the appropriate output responses.

Problem 6.6. Develop an ILQR theory for systems with actuator quantization. Specifically, consider the SISO LPNI system specified by

$$\dot{x} = Ax + B_1 qz(u) + B_2 d,$$
$$y = C_1 x,$$
$$u = Kx,$$

where $x \in \mathbf{R}^{n_x}$ and $qz(\cdot)$ is given by (2.35). Using the definition of $Q_m(\cdot)$ in (6.125) and assuming that $Q_m^{-1}(\cdot)$ exists, find a closed form solution for the ILQR problem

$$\min_{K,\Delta} \left(\sigma_y^2 + \rho \sigma_u^2 + \eta \frac{1}{\Delta^2} \right), \rho > 0, \eta > 0.$$

Problem 6.7. Develop an ILQG theory for systems with sensor quantization. Specifically, consider the SISO LPNI system specified by

$$\dot{x} = Ax + B_1 u + B_2 d,$$
$$y = C_1 x,$$
$$y_m = qz_\Delta(y),$$

where $x \in \mathbf{R}^{n_x}$ and $qz(\cdot)$ is given by (2.35). The controller is specified by

$$\dot{x}_c = Mx_c - Ly_m,$$
$$u = Kx_c,$$

where $x_c \in \mathbf{R}^{n_{x_c}}$. Using the definition of $Q_m(\cdot)$ in (6.125) and assuming that $Q_m^{-1}(\cdot)$ exists, find a closed form solution for the ILQG problem

$$\min_{K,\Delta} \left(\sigma_y^2 + \rho \sigma_u^2 + \eta \frac{1}{\Delta^2} \right), \rho > 0, \eta > 0.$$

Problem 6.8. Using the solution of Problem 6.7, repeat parts (c) and (d) of Problem 6.5 but with a quantized measurement, that is, $y_m = \text{qz}_\Delta(y)$. Here, assume that the actuator is linear (i.e., the voltage does not saturate).

Problem 6.9. Consider the active suspension of Problem 6.1, where the disturbance d is standard white noise. Assume that the actuator is linear and the outputs are $y_1 = x_1 - x_2$ and $y_2 = v_1 - v_2$. Moreover, assume that the sensors are saturating, that is, $y_{m_1} = \text{sat}_{\alpha_1}(y_1)$ and $y_{m_2} = \text{sat}_{\alpha_2}(y_2)$. Use the MIMO extension technique of Subsection 6.2.4 to find an ILQG solution that ensures 98% disturbance rejection. What are the required values of α_1 and α_2?

Problem 6.10. The ILQR problem considered in Section 6.2 is formulated in terms of disturbance rejection. Establish the duality of this problem with the ILQR problem for reference tracking. To accomplish this, consider the SISO system

$$\dot{x} = Ax + B_1\text{sat}_\alpha(u),$$
$$y = C_1x,$$
$$u = r - Kx,$$

where $x \in \mathbf{R}^{n_x}$ and the reference r is generated by the system

$$\dot{x}_r = A_r x_r + B_2 w,$$
$$r = C_2 x_r,$$

where $x_r \in \mathbf{R}^{n_{x_r}}$. Find a closed form solution for the ILQR problem

$$\min_{K,\alpha} \left(\sigma_e^2 + \rho\sigma_u^2 + \eta\alpha^2 \right), \rho > 0, \eta > 0,$$

where $e = y - r$.

Problem 6.11. Prove that it is impossible to obtain multiple solutions of stochastic linearization equation (2.52) when using an SLQR or ILQR controller.

6.5 Annotated Bibliography

The origin of LQR/LQG theory can be found in the following:

[6.1] R.E. Kalman, "Contributions to the theory of optimal control," *Boletin de la Sociedad Matematica Mexicana*, Vol. 5, pp. 102–119, 1960

[6.2] R.E. Kalman, "A new approach to linear filtering and prediction problems," *Transactions of the ASME – Journal of Basic Engineering*, Vol. 82 (Series D), pp. 35–45, 1960

[6.3] A.M. Letov, "Analytic construction of regulators, I" *Automation and Remote Control*, Vol. 21, No. 4, pp. 436–441, 1960

Subsequent developments have been summarized in the following:

[6.4] H. Kwakernaak and R. Sivan, *Linear Optimal Control Systems*, Wiley-Interscience, New York, 1972

[6.5] B.D.O. Anderson and J.B. Moore, *Optimal Control: Linear Quadratic Methods*, Prentice Hall, Englewood Cliffs, NJ, 1989

The material of Section 6.1 is based on the following:

[6.6] C. Gokcek, P.T. Kabamba, and S.M. Meerkov, "An LQR/LQG theory for systems with saturating actuators," *IEEE Transactions on Automatic Control*, Vol. 46, No. 10, pp. 1529–1542, 2001

A description of ship roll stabilization model can be found in the following:

[6.7] R. Bhattacharyya, *Dynamics of Marine Vehicles*, John Wiley and Sons, New York, 1978

[6.8] T.I. Fossen, *Guidance and Control of Ocean Vehicles*, John Wiley and Sons, New York, 1994

[6.9] L. Fortuna and G. Muscato, "A roll stabilization systems for a monohull ship: Modeling, identification and adaptive control," *IEEE Transactions on Control Systems Technology*, Vol. 4, No. 1, pp. 18–28, 1996

The material of Section 6.2 is based on the following:

[6.10] S. Ching, P.T. Kabamba, and S.M. Meerkov, "Instrumented LQR/LQG: A method for simultaneous design of controller and instrumentation," *IEEE Transactions on Automatic Control*, Vol. 55, No. 1, pp. 217–221, 2010

7 Performance Recovery in LPNI Systems

Motivation: The nonlinearities in sensors and actuators lead to performance degradation in LPNI systems as compared with linear ones. In this situation, two questions arise: (1) How can nonlinear instrumentation be selected so that the degradation is no more than a given bound? (2) Is it possible to modify the controller so that for a given nonlinear instrumentation, the linear performance is recovered? These questions motivate the subject matter of this chapter. Specifically, the first question leads to a method of partial performance recovery and the second to a complete performance recovery technique.

Overview: We develop techniques for both partial and complete performance recovery. Specifically, we show that the solution of the former is given by a Nyquist stability criterion-type technique (Section 7.1), while the latter is provided by the so-called controller boosting approach (Section 7.2).

7.1 Partial Performance Recovery

7.1.1 Scenario

In practice, control systems are often designed using linear techniques. In reality, control systems often (or, perhaps, always) include saturating actuators. The question arises: How large should the level of saturation be so that the performance predicted by linear design does not degrade too much? In this section, this question is addressed in the framework of the disturbance rejection problem. The scenario is as follows:

- The disturbance is a Gaussian random process and the performance is measured by the standard deviation of the output.
- The linear design (i.e., the design under the assumption that no saturation takes place) results in the output standard deviation, denoted by σ_{y_l}.
- The saturation is described by the usual expression

$$
\mathrm{sat}_\alpha(u) = \begin{cases} \alpha, & u > \alpha, \\ u, & -\alpha \leq u \leq \alpha, \\ -\alpha, & u < -\alpha, \end{cases} \tag{7.1}
$$

where α is the level of saturation.

- The tolerable performance degradation is characterized by a number $e > 0$, that is,

$$\sigma_y \leq (1+e)\sigma_{y_l}, \qquad (7.2)$$

where σ_y is the standard deviation of the output, y, of the closed loop system with the controller obtained by the above-mentioned linear design and with saturating actuator (7.1).

Within this scenario, we provide a method for selecting the level of actuator saturation, α, which results in performance degradation no more than e.

7.1.2 Problem Formulation

Consider the SISO linear system shown in Figure 7.1(a), where $P(s)$ is the plant, $C(s)$ the controller, $F(s)$ the coloring filter, and w, u_l, and y_l, are standard white noise, control signal, and output, respectively. Suppose that $C(s)$ is designed using a linear design technique so that the system of Figure 7.1(a) is asymptotically stable and transfer functions $\frac{FP}{1+PC}$ and $\frac{FPC}{1+PC}$ are strictly proper. Assume also that the goal of control is disturbance rejection, quantified by the standard deviation of y_l. Under these assumptions, it is easy to see that

$$\sigma_{y_l} = \left\| \frac{F(s)P(s)}{1+P(s)C(s)} \right\|_2, \quad \sigma_{u_l} = \left\| \frac{F(s)P(s)C(s)}{1+P(s)C(s)} \right\|_2; \qquad (7.3)$$

Assume now that the above controller is used for the same plant but with a saturating actuator. The resulting LPNI system is shown in Figure 7.1(b), where all

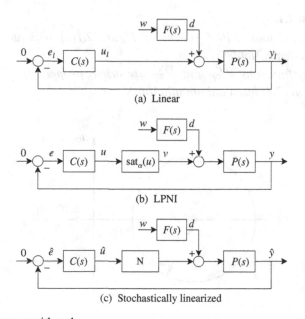

(a) Linear

(b) LPNI

(c) Stochastically linearized

Figure 7.1. Systems considered.

signals are denoted by the same symbols as in Figure 7.1(a) but without the subscript "l". Assume further that this system remains globally asymptotically stable. One would like to compare the standard deviation of y_l with that of y. To accomplish this, we use the method of stochastic linearization. The stochastically linearized system is shown in Figure 7.1(c), where N satisfies

$$N = \mathrm{erf}\left(\frac{\alpha}{\sqrt{2}\left\|\frac{F(s)P(s)C(s)}{1+NP(s)C(s)}\right\|_2}\right). \tag{7.4}$$

Clearly, the standard deviation of $\hat{y}$ of Figure 7.1 is given by

$$\sigma_{\hat{y}} = \left\|\frac{F(s)P(s)}{1+NP(s)C(s)}\right\|_2. \tag{7.5}$$

Assume that the acceptable level of performance degradation is defined by

$$\sigma_{\hat{y}} \leq (1+e)\sigma_{y_l}, \tag{7.6}$$

where $e > 0$. The problem considered in this section is: *Given $e > 0$, find a lower bound on the level of saturation, α, so that $\sigma_{\hat{y}} \leq (1+e)\sigma_{y_l}$.*

7.1.3 Main Result

Let $D(r)$ denote a closed disk in the complex plane with radius r, centered at $(-r-1, j0)$ (see Figure 7.2).

Introduce the following assumptions:

Assumption 7.1.
 (i) *The closed loop LPNI system of Figure 7.1(a) with $w = 0$ is globally asymptotically stable;*
 (ii) *Transfer functions $\frac{FP}{1+PC}$ and $\frac{FPC}{1+PC}$ are strictly proper;*
 (iii) *Equation (7.4) has a unique solution N.*

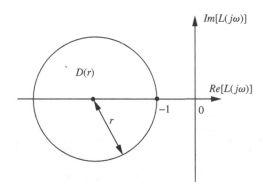

Figure 7.2. Disk $D(r)$.

Theorem 7.1. *Let Assumption 7.1 hold, and r be such that the Nyquist plot of the loop gain $P(s)C(s)$ lies entirely outside of $D(r)$. Then, for any $e > 0$,*

$$\sigma_{\hat{y}} \leq (1+e)\sigma_{y_l} \tag{7.7}$$

if

$$\alpha > \beta(e,r)\sigma_{u_l}, \tag{7.8}$$

where

$$\beta(e,r) = \sqrt{2}(1+e)\text{erf}^{-1}\left[\frac{2r+(1+e)}{(1+e)(2r+1)}\right]. \tag{7.9}$$

Proof. See Section 8.6.

Figure 7.3 illustrates the behavior of $\beta(e,r)$ with $e = 0.1$ and $e = 0.05$, for a wide range of r. Obviously, $\beta(e,r)$ is not very sensitive to r for $r > 1$. In particular, it is close to 2 for all $r > 1$, if $e = 0.1$. Based on this we formulate:

Rule of thumb 7.1. If $\alpha \geq 2\sigma_{u_l}$, the performance degradation of the linear design due to actuator saturation is less then 10%.

7.1.4 Examples

Example 7.1. Consider the feedback system with P-controller shown in Figure 7.4, where w is standard white noise. Using the Popov criterion, one can easily check that this system is globally asymptotically stable for all $\alpha > 0$. If no saturation takes place, $\sigma_{y_l} = 0.02357$ and $\sigma_{u_l} - 1.4142$. To select a level of saturation, α, that results

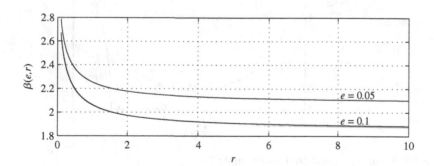

Figure 7.3. Function $\beta(e,r)$.

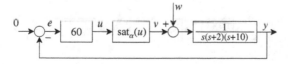

Figure 7.4. Example 7.1.

in less than 10% performance degradation, we draw the Nyquist plot of $P(s)C(s) = \frac{60}{s(s+2)(s+10)}$ and determine the largest disk $D(r)$ such that $P(j\omega)C(j\omega)$ lies entirely in its exterior. It turns out that $r = 4.2$, as shown in Figure 7.5.

Thus, according to Theorem 7.1,

$$\alpha \geq \beta(0.1, 4.2)\sigma_{u_l} = (1.9)(1.4142) = 2.687 \tag{7.10}$$

guarantees that the degradation of performance is at most 10%. With $\alpha = 2.687$, the standard deviation of the output is $\sigma_{\hat{y}} = 0.0241$, which is larger than σ_{y_l} by 2.2%. To obtain σ_y, we simulate the system of Figure 7.4 and evaluate σ_y to be 0.0242. This implies that the performance degradation of the LPNI system is 2.7% and the error of stochastic linearization in predicting σ_y is 0.4%.

Example 7.2. Consider the feedback control system with *PI*-controller shown in Figure 7.6. Here again, using the Popov criterion, one can easily check that this system is globally asymptotically stable. Without the saturation, $\sigma_{y_l} = 0.03571$ and $\sigma_{u_l} = 0.4432$. Again, assume that 10% degradation of σ_y from σ_{y_l} is acceptable. The Nyquist plot of $P(s)C(s) = \frac{11s+6}{s(s^2+2s+6)}$ lies entirely outside of the disk $D(80)$. Therefore, $\alpha \geq \beta(0.1, 80)\sigma_{u_l} = 1.86 \times 0.4432 = 0.8244$ achieves the desired performance. With $\alpha = 0.8244$, $\sigma_{\hat{y}} = 0.03723$, which is larger than σ_{y_l} by 4.3%. The results of simulations

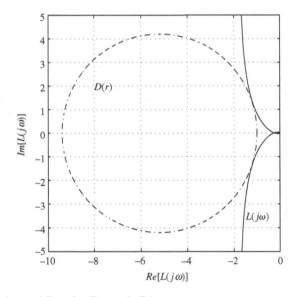

Figure 7.5. Nyquist plot and $D(r)$ for Example 7.1.

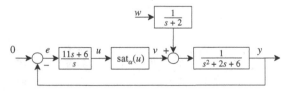

Figure 7.6. Example 7.2.

provide $\sigma_y = 0.0371$. The performance degradation of the LPNI system is 3.9% and the error of stochastic linearization in predicting σ_y is 0.35%.

7.2 Complete Performance Recovery

7.2.1 Scenario

Consider the linear feedback system shown in Figure 7.7(a), where $C(s)$ is the controller and $P(s)$ is the plant. The signals u_l, y_l, and w denote the controller output, plant output, and standard white input disturbance, respectively. The disturbance here is not filtered to simplify the presentation; an extension to a disturbance filtered by $F_\Omega(s)$ is straightforward. Assume that the controller, $C(s)$, is designed to achieve a certain level of disturbance rejection, specified in terms of the output standard

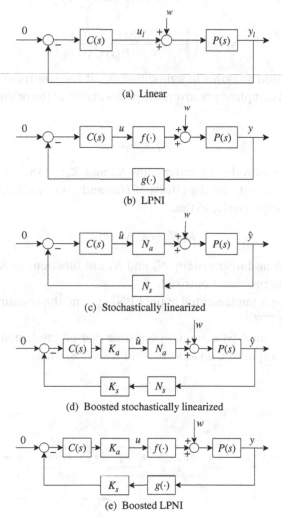

Figure 7.7. Systems considered.

deviation σ_{y_l}. In reality, the controller is implemented in the LPNI configuration shown in Figure 7.7(b), where $f(\cdot)$ and $g(\cdot)$ represent odd static nonlinearities in the actuator and sensor, respectively.

The performance of the LPNI system typically degrades in comparison with that of the original linear system in the sense that

$$\sigma_y > \sigma_{y_l}. \tag{7.11}$$

This section presents a technique, referred to as *boosting*, that describes how, under certain conditions, the gain of $C(s)$ can be increased to eliminate this degradation.

To accomplish this, we again use the method of stochastic linearization. The stochastically linearized version of the system of Figure 7.7(b) is shown in Figure 7.7(c), where N_a and N_s are solutions of the following equations:

$$N_a = \mathcal{F}\left(\left\|\frac{P(s)N_sC(s)}{1 + N_aN_sP(s)C(s)}\right\|_2\right), \tag{7.12}$$

$$N_s = \mathcal{G}\left(\left\|\frac{P(s)}{1 + N_aN_sP(s)C(s)}\right\|_2\right). \tag{7.13}$$

The boosting method amounts to a modification of the controller $C(s)$ so that the quasilinear system completely recovers the performance of the original linear system, that is,

$$\sigma_{\hat{y}} = \sigma_{y_l}. \tag{7.14}$$

This is achieved by introducing scalar gains K_a and K_s, as shown in Figure 7.7(d). The idea is to compensate for the effects of $f(u)$ and $g(y)$ by selecting K_a and K_s to offset N_a and N_s respectively, so that

$$K_aN_a = K_sN_s = 1. \tag{7.15}$$

Note that, in the quasilinear system, N_a and N_s are functions of K_a and K_s, which makes the boosting problem nontrivial.

When boosting is implemented in the LPNI system, the resulting block diagram is shown in Figure 7.7(d).

Since K_s, K_a, and $C(s)$ commute, boosting can be implemented by placing a single gain at the output of $C(s)$ as shown in Figure 7.8, where

$$K_{boost} := K_aK_s. \tag{7.16}$$

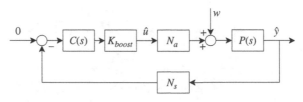

Figure 7.8. Equivalent boosted quasilinear system.

In this section, we derive conditions under which K_a and K_s exist and provide a method for their calculation. In addition, we establish a separation principle, which enables K_a and K_s to be evaluated from the following two simpler subproblems:

(i) *a*-boosting, that is, boosting to account for only a nonlinear actuator (i.e., under the assumption that $g(y) = y$), and

(ii) *s*-boosting, that is, boosting to account for only a nonlinear sensor (i.e., under the assumption that $f(u) = u$).

7.2.2 Problem Formulation

a-Boosting: Consider the LPNI system of Figure 7.7(b) with $g(y) = y$. Hence, the only nonlinearity is the actuator $f(u)$, and Figure 7.7(d) reduces to Figure 7.9, where

$$N_a = \mathcal{F}\left(\left\|\frac{P(s)C(s)K_a}{1+P(s)N_aK_aC(s)}\right\|_2\right) \tag{7.17}$$

and

$$\mathcal{F}(\sigma) := \int_{-\infty}^{\infty} f'(x)\frac{1}{\sigma\sqrt{2\pi}}\exp\left(-\frac{x^2}{2\sigma^2}\right)dx. \tag{7.18}$$

The problem of a-boosting is: *Find, if possible, K_a such that*

$$K_aN_a = 1, \tag{7.19}$$

where N_a itself depends on K_a through (7.17).

s-Boosting: Consider the LPNI system with $f(u) = u$. Hence, Figure 7.7(b) reduces to Figure 7.10, where

$$N_s = \mathcal{G}\left(\left\|\frac{P(s)}{1+P(s)N_sK_sC(s)}\right\|_2\right) \tag{7.20}$$

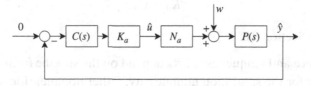

Figure 7.9. *a*-Boosted quasilinear system.

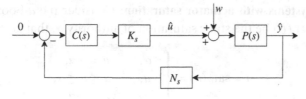

Figure 7.10. *s*-Boosted quasilinear system.

and

$$\mathcal{G}(\sigma) := \int_{-\infty}^{\infty} g'(x) \frac{1}{\sigma\sqrt{2\pi}} \exp\left(-\frac{x^2}{2\sigma^2}\right) dx. \tag{7.21}$$

The problem of s-boosting is: *Find, if possible, K_s such that*

$$K_s N_s = 1, \tag{7.22}$$

where, again, N_s is a function of K_s through (7.20).

The structure of the LPNI system of Figure 7.7(e) implies that the problems of a- and s-boosting are not dual. Indeed, observe that for a-boosting, the gain K_a appears in the forward path between w and the input of the actuator nonlinearity, $\hat{u}$. For s-boosting, K_s does not appear in the path from w to the input of the sensor $\hat{y}$. Consequently, the numerator of the transfer function in (7.17) contains a boosting gain, whereas that in (7.20) does not. Thus, the two problems are different, and must be addressed separately.

7.2.3 a-Boosting

Necessary and sufficient condition: As implied by (7.17) and (7.19), the problem of a-boosting is equivalent to finding K_a that satisfies

$$K_a \mathcal{F}\left(\left\|\frac{P(s)C(s)K_a}{1+P(s)N_a K_a C(s)}\right\|_2\right) = 1. \tag{7.23}$$

Theorem 7.2. *a-Boosting is possible if and only if*

$$x\mathcal{F}\left(x\left\|\frac{P(s)C(s)}{1+P(s)C(s)}\right\|_2\right) = 1 \tag{7.24}$$

has a positive solution. Any positive solution, x^*, of (7.24) yields a boosting gain

$$K_a = x^*. \tag{7.25}$$

Proof. See Section 8.6.

The existence and uniqueness of K_a depend on the specific form of $\mathcal{F}(\cdot)$. This is analyzed below for the saturation nonlinearity. Other nonlinearities can be treated analogously.

a-Boosting in systems with actuator saturation: Consider the a-boosted system of Figure 7.9 and let $f(\cdot)$ be the static saturation of authority α, that is,

$$f(u) = \mathrm{sat}_\alpha(u) = \begin{cases} \alpha, & u > +\alpha, \\ u, & -\alpha \le u \le \alpha, \\ -\alpha, & u < -\alpha. \end{cases} \tag{7.26}$$

In this case,

$$N_a = \text{erf}\left(\frac{\alpha}{\sqrt{2}\left\|\frac{P(s)C(s)K_a}{1+P(s)N_aK_aC(s)}\right\|_2}\right). \tag{7.27}$$

It follows from Theorem 7.2 that a-boosting for the saturation nonlinearity (7.26) is possible if and only if the equation

$$x\text{erf}\left(\frac{c}{x}\right) = 1 \tag{7.28}$$

has a positive solution, where

$$c = \frac{\alpha}{\sqrt{2}\left\|\frac{P(s)C(s)}{1+P(s)C(s)}\right\|_2}. \tag{7.29}$$

Theorem 7.3. *Equation (7.28) admits a unique positive solution if and only if*

$$\alpha > \sqrt{\frac{\pi}{2}}\left\|\frac{P(s)C(s)}{1+P(s)C(s)}\right\|_2. \tag{7.30}$$

Proof. See Section 8.6.

Note that since

$$\sigma_{u_l} = \left\|\frac{P(s)C(s)}{1+P(s)C(s)}\right\|_2 \tag{7.31}$$

and

$$\sqrt{\frac{\pi}{2}} \approx 1.25, \tag{7.32}$$

the following can be stated:

Rule of thumb 7.2. a-Boosting for a saturating actuator is possible if

$$\alpha > 1.25\sigma_{u_l}. \tag{7.33}$$

Recall that, as it has been shown in Section 7.1,

$$\alpha > 2\sigma_{u_l}, \tag{7.34}$$

without boosting, leads to no more than 10% performance degradation of the linear design. In comparison, the above rule of thumb implies that complete performance recovery is possible even if the actuator has less authority than recommended in Section 7.1.

Performance recovery by redesigning $C(s)$**:** If (7.24) does not have a solution, that is, a-boosting of $C(s)$ is impossible, a question arises: Can $C(s)$ be redesigned in some other way to achieve $\sigma_{\hat{y}} = \sigma_{y_l}$? The answer depends on the ability to find a controller that simultaneously achieves the performance specification with linear actuator *and* yields a solution to (7.24). Such a controller is said to be *boostable*.

In the case of actuator saturation, the boosting condition (7.30) implies that finding a boostable controller is a linear minimum-effort control problem, that is, the problem of finding a controller, $C_{opt}(s)$, that minimizes σ_{u_l} for a specified performance level σ_{y_l}. If (7.30) is not satisfied by this $C_{opt}(s)$, then no linear boostable controller exists.

7.2.4 s-Boosting

Boosting equation: As implied by (7.20) and (7.22), the problem of s-boosting is equivalent to finding K_s that satisfies

$$K_s \mathcal{G} \left(\left\| \frac{P(s)}{1 + P(s) N_s K_s C(s)} \right\|_2 \right) = 1. \tag{7.35}$$

Since, unlike a-boosting, K_s enters the argument of $\mathcal{G}$ only as a factor of N_s, and for s-boosting $N_s K_s = 1$, the solution of (7.35) is always possible and is given by

$$K_s = \frac{1}{\mathcal{G} \left(\left\| \frac{P(s)}{1 + P(s)C(s)} \right\|_2 \right)}. \tag{7.36}$$

This result warrants further consideration since it suggests that linear performance may be recovered in the presence of *any* sensor nonlinearity. It turns out that, although an s-boosting gain can always be found, in some cases the accuracy of stochastic linearization may be poor. Thus, certain conditions should be satisfied before using s-boosting. These are developed in Subsection 7.2.7.

7.2.5 Simultaneous a- and s-Boosting

The following separation principle ensures that the results of Subsections 7.2.3 and 7.2.4 remain applicable when actuator and sensor nonlinearities are simultaneously present.

Theorem 7.4. *Simultaneous a- and s-boosting is possible if and only if each is possible independently. Moreover, the boosting gains, K_a and K_s, are the same as the individual a- and s-boosting gains, respectively.*

Proof. See Section 8.6.

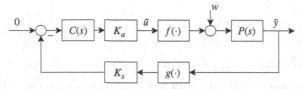

Figure 7.11. LPNI system with boosted controller for stability verification.

7.2.6 Stability Verification in the Problem of Boosting

It is important to verify the stability of the original LPNI system with the boosted controller. This amounts to performing stability analysis for the nonlinear system shown in Figure 7.11.

The stability properties of the LPNI system can be ascertained using the theory of absolute stability. Here, the global asymptotic stability of the origin can be verified by using tools such as the circle or Popov criteria. Although these are strong results, the conditions are only sufficient and not necessary.

In general, for the LPNI system of Figure 7.11, the local stability of any equilibrium can be ascertained via Lyapunov's indirect method. For example, in the case where $f(\cdot)$ and $g(\cdot)$ are saturation functions of the form (1.8), the origin is asymptotically stable if the numerator of the equation

$$1 + K_a K_s C(s) P(s) = 0$$

has all roots in the open left half plane. In this case, the region of attraction can be estimated by using well-known Lyapunov function-based techniques.

7.2.7 Accuracy of Stochastic Linearization in the Problem of Boosting

It is assumed that the boosted quasilinear system accurately predicts the behavior of the corresponding nonlinear system. This section validates the accuracy of stochastic linearization when boosting is performed. Design guidelines are formulated to avoid cases where accuracy is poor.

Accuracy of a-boosting: To validate the accuracy of stochastic linearization in the context of a-boosting, the following statistical study, similar to that described in Subsection 2.3.2, is performed: We consider 2500 first-order and 2500 second-order plants of the form:

$$P_1(s) = \frac{1}{Ts+1},$$
(7.37)

$$P_2(s) = \frac{\omega_n^2}{s^2 + 2\zeta\omega_n + \omega_n^2}.$$
(7.38)

The controller is $C(s) = K$ and the actuator is a saturation of the form (7.26). The system parameters are randomly and equiprobably selected from the following sets:

$T \in [0.01, 10]$,
$\omega_n \in [0.01, 10]$, $\zeta \in [0.05, 1]$,
$K \in [1, 20]$,
$\alpha \in (\alpha_{min}, 2\alpha_{min}]$,

where α_{min} is the right-hand side of (7.30). Boosting is performed for each system, and the LPNI system is simulated to identify the error of stochastic linearization, defined as

$$e_{SL} = \frac{|\sigma_y - \sigma_{\hat{y}}|}{\sigma_{\hat{y}}}. \tag{7.39}$$

Accuracy is very good: 73.6% of the systems yield $e_{SL} < 0.05$ and only 8.7% of systems yield $e_{SL} > 0.1$. Further analysis reveals that these latter cases occur when the signals u and y are highly non-Gaussian. This is consistent with the assumption of stochastic linearization, namely that those signals should be approximately Gaussian.

In general, stochastic linearization is accurate when the closed loop linear system provides a sufficient amount of low-pass filtering. A similar situation holds for the method of describing functions.

Accuracy of s-boosting: A similar statistical study is performed to validate the accuracy of stochastic linearization in the context of s-boosting. Here, the sensor is assumed to be a quantizer with saturation given in (2.38), and $C(s) = K$. We consider 1000 first- and 1000 second-order plants of the form (7.37) and (7.38), with system parameters chosen equiprobably from the following sets:

$T \in [0.01, 10]$,
$\omega_n \in [0.01, 10]$, $\zeta \in [0.05, 1]$,
$K \in [1, 20]$,
$m \in [1, 10]$, $\Delta \in (0, 2\sigma_{y_l}]$,

where σ_{y_l} is the nominal linear performance to be recovered. Let $\Delta_1 = 2\Delta$ denote the quantizer deadzone. As illustrated in Figure 7.12, simulations reveal that accuracy degrades significantly as the ratio σ_{y_l}/Δ_1 decreases. This is expected, since when σ_{y_l}/Δ_1 is small, most of the output signal lies in the quantizer deadzone. Hence, the nonlinear system operates in an effectively open loop regime. Our experience indicates that to avoid this situation, the following should be observed:

$$\frac{\sigma_{y_l}}{\Delta_1} > 0.33. \tag{7.40}$$

This leads to:

Rule of thumb 7.3. s-Boosting for a quantized sensor is possible if $\Delta_1 < 3\sigma_{y_l}$, that is,

$$\Delta < 1.5\sigma_{y_l}. \tag{7.41}$$

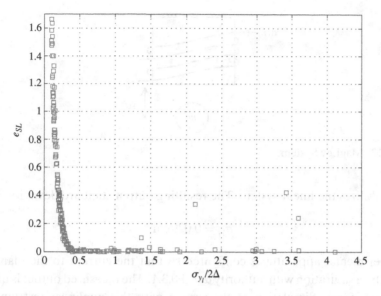

Figure 7.12. e_{SL} as function of $\sigma_{y_l}/2\Delta$.

This rule of thumb may seem "generous," since intuition would suggest that Δ should be, at most, one standard deviation. The extra deadzone width allowance comes from boosting, which increases the loop gain.

When (7.41) is satisfied, the accuracy of s-boosting is similar to that of a-boosting. Again, accuracy is generally very good, and fails in those scenarios where the plant has insufficient filtering characteristics.

Similar results hold when $g(y)$ is the deadzone nonlinearity (2.23). In general, Rule of thumb 7.3 should be observed for any sensor nonlinearities that exhibit small gain near the origin.

7.2.8 Application: MagLev

Consider the problem of controlling the vertical displacement of a magnetically suspended ball (MagLev), illustrated in Figure 7.13. The input of the system is the current $i(t)$, while the output is the vertical displacement of the ball $y(t)$.

To coincide with a commercial MagLev experimental apparatus (Feedback Inc. Magnetic Levitation System), the following linearized model and the disturbance intensity, provided by the manufacturer, are used:

$$P(s) = \frac{Y(s)}{I(s)} = \frac{37.20}{s^2 - 2180}, \tag{7.42}$$

$$\sigma_d^2 = 1.2 \times 10^{-5}. \tag{7.43}$$

Consider the PID controller

$$C(s) = 200 + 5s + \frac{200}{s}, \tag{7.44}$$

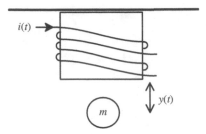

Figure 7.13. MagLev System.

which stabilizes the linear system. The resulting output standard deviation is

$$\sigma_{y_l} = 0.029 \, \text{cm}. \tag{7.45}$$

The experimental apparatus is configured so that the current to the plant is constrained by a saturation with authority $\alpha = \pm 0.3A$. The measured output is quantized by $\Delta = 0.005$ cm. Simulation of the system with this nonlinear instrumentation results in

$$\sigma_y = 0.049 \, \text{cm}, \tag{7.46}$$

a degradation of 83%.

It is easily verified that $\sigma_{u_l} = 0.1069$, and hence (7.30) is satisfied. Thus, the conditions of Theorem 7.4 are met, and boosting can be used to recover the original linear performance. Solving (7.28) and (7.36) results in

$$K_a K_s = K_{boost} = 2.8. \tag{7.47}$$

Using this boosting gain in a MATLAB simulation yields the desired result:

$$\sigma_y = 0.0271 \, \text{cm}. \tag{7.48}$$

This result is verified experimentally. The boosted controller $\bar{C}(s) = K_{boost} C(s)$ is applied to the MagLev through standard AD/DA hardware, and a pseudo-white noise excitation is applied at the input of the plant. The resulting experimentally measured output standard deviation is

$$\sigma_{\tilde{y}} = 0.0252 \, \text{cm}, \tag{7.49}$$

where $\tilde{y}$ denotes the output signal measured experimentally. Thus, a successful recovery of the designed performance is demonstrated.

7.3 Summary

- The problem of partial recovery of linear performance is solved using the Nyquist plot of the loop gain.

- Based on this solution, the following rule of thumb is obtained: The performance degradation is less then 10% if

$$\alpha > 2\sigma_{u_l},$$

where α is the level of actuator saturation and σ_{u_l} is the standard deviation of the signal at the output of the controller in the case of a linear actuator.
- The complete recovery of linear performance can be attained by boosting controller gains to account for nonlinearities in actuators (a-boosting) and sensors (s-boosting).
- A necessary and sufficient condition for a-boosting is provided in terms of positive solutions of a transcendental equation.
- The following rule of thumb stems from this equation: a-Boosting for saturating actuators is possible if

$$\alpha > 1.25\sigma_{u_l},$$

where, as before, α is the level of actuator saturation and σ_{u_l} is the standard deviation of the signal at the output of the controller in the case of a linear actuator.
- s-Boosting is always possible, as long as stochastic linearization can be applied. In the case of a quantized sensor, this occurs when

$$\Delta < 1.5\sigma_{y_l}$$

where Δ is the quantization step and σ_{y_l} is the standard deviation of the system output in the case of a linear sensor.
- For the boosting method, a separation principle holds: a- and s-boosting designed separately provide complete boosting of an LPNI system when they are applied simultaneously.
- The methods of this chapter provide solutions for the Partial and Complete Performance Recovery problems introduced in Section 1.2.

7.4 Problems

Problem 7.1. Consider the linear system shown in Figure 7.14 and assume that

$$P(s) = \frac{1}{s(s+2)}, \quad C(s) = 3, \quad F(s) = \frac{1}{s+0.1}. \tag{7.50}$$

(a) Find the standard deviation of y_l.
(b) Draw the Nyquist plot of the loop gain $L(s) = P(s)C(s)$.
(c) Find the largest r such that the Nyquist plot lies entirely outside of the disk $D(r)$, where $D(r)$ is defined in Figure 7.2.
(d) Assume now that the actuator is saturating. Use Theorem 7.1 to select the actuator saturation level α so that performance degradation is not more than 5%.

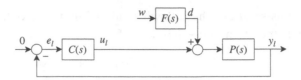

Figure 7.14. Problem 7.1.

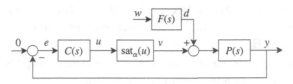

Figure 7.15. Problem 7.2.

(e) Use the rule of thumb of Subsection 7.1.3 to select the actuator saturation level α so that performance degradation is not more than 5%. Compare the resulting saturation level with that from part (d).

Problem 7.2. Consider the linear system shown in Figure 7.4 and assume that

$$P(s) = \frac{1}{s}, \quad C(s) = 10, \quad F(s) = \frac{0.3}{s+0.1}. \tag{7.51}$$

(a) Use Theorem 7.1 to select the actuator saturation level α so that performance degradation from linear design is not more than 10%.
(b) Use the rule of thumb of Subsection 7.1.3 to select the actuator saturation level α so that performance degradation is not more than 10%.
(c) Plot the standard deviation of y as a function of α and from the plot select α so that performance degradation is not more than 10%. Compare the result with those from part (a) and (b).

Problem 7.3. Repeat parts (a), (b), and (c) of Problem 7.2 for the system of Figure 7.15, where

$$P(s) = \frac{1.484 \times 10^{-6}}{s^2 + 1.125s + 1.563}, \quad C(s) = \frac{132.36(s+0.1)}{(s+10)(s+1.57)},$$

$$F(s) = \frac{7.224 \times 10^4 s}{s^2 + 0.286s + 0.311}. \tag{7.52}$$

Problem 7.4. Consider the linear system in Figure 7.14 and assume that

$$P(s) = \frac{1}{s+1}, \quad F(s) = \frac{1}{5s+1}. \tag{7.53}$$

(a) Design $C(s)$ so that the standard deviation of y_l is less or equal to 0.1.

(b) Assume now that the actuator is saturating and select α so that performance
degradation is not more than 5%.

Problem 7.5. Consider the disturbance rejection system of Figure 7.7(a). Assume
that the plant and controller are given by

$$P(s) = \frac{1}{s(s+2)(s+4)}, \quad C(s) = 1. \tag{7.54}$$

(a) Calculate the standard deviation of the output, σ_{y_l}.
(b) Now, assume that the actuator is a saturation with $\alpha = 0.5$ and the sensor is
linear (see Figure 7.7(b)). By simulation, determine σ_y and assess the level
of performance degradation as compared with σ_{y_l}.
(c) Calculate the a-boosting gain for performance recovery.
(d) By simulation, determine σ_y when the controller has been augmented with
the a-boosting gain. Has the performance been recovered?

Problem 7.6. Consider the system of Figure 7.7(a). Assume that the plant and
controller are given by

$$P(s) = \frac{1}{s(s+1)(s+2)}, \quad C(s) = 1, \tag{7.55}$$

resulting in a desirable level of disturbance rejection σ_{y_l}. Now, assume that the
system is implemented in the configuration of Figure 7.7(b), where the actuator is
the saturation $f(\cdot) = \mathrm{sat}_\alpha(\cdot)$ and the sensor is linear.

(a) Determine the range of α for which a-boosting is possible.
(b) Plot the a-boosting gain as a function of α.
(c) Now, assume that the actuator is the deadzone $\mathrm{dz}_\Delta(\cdot)$ defined in (2.23).
Determine the range of Δ for which a-boosting is possible.
(d) Plot the a-boosting gain as a function of Δ.

Problem 7.7. Find conditions for a-boosting in the case of reference tracking.
Specifically, consider the system of Figure 7.16(a), where $P(s)$ is the plant and $C(s)$

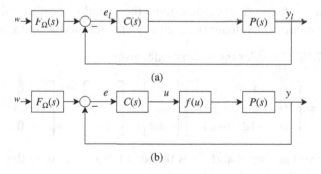

(a)

(b)

Figure 7.16. Problem 7.7.

is the controller designed to achieve a standard deviation of tracking error σ_{e_l}. Now, assume that the system is implemented with a nonlinear actuator as shown in Figure 7.16(b), resulting in performance degradation.

(a) Find an expression for an a-boosting gain that recovers the linear performance, that is, $\sigma_{\hat{e}} = \sigma_{e_l}$.
(b) Does this boosting gain always exist?
(c) By simulation, demonstrate the accuracy of a-boosting in reference tracking using the plant and controller

$$P(s) = \frac{1}{s(s+1)}, \ C(s) = 1.$$

Problem 7.8. Consider the system of Figure 7.7(a). Assume that the plant and controller are given by

$$P(s) = \frac{1}{s(s+2)}, \ C(s) = 1, \tag{7.56}$$

resulting in a desirable level of disturbance rejection σ_{y_l}. Now, assume that the system is implemented in the configuration of Figure 7.7(b), where the actuator is linear and the sensor is the quantization $g(\cdot) = \text{qn}_\Delta(\cdot)$, given by (2.35).

(a) Plot the s-boosting gain as a function of Δ.
(b) By simulation, show that the accuracy of s-boosting deteriorates as Δ increases.

Problem 7.9. Consider the system of Figure 7.7(a). Assume that the plant and controller are given by

$$P(s) = \frac{1}{s(s+2)}, \ C(s) = 1, \tag{7.57}$$

resulting in a desirable level of disturbance rejection σ_{y_l}. Now, assume that the system is implemented in the configuration of Figure 7.7(b), where the actuator is the saturation $f(\cdot) = \text{sat}_\alpha(\cdot)$ and the sensor is the deadzone $g(\cdot) = \text{dz}_\Delta(\cdot)$ given by (2.23).

(a) For $\alpha = 0.5$ and $\Delta = 0.2$, determine the a- and s-boosting gains.
(b) By simulation, evaluate the accuracy of combined a- and s-boosting.

Problem 7.10. Consider the servo model given by

$$\begin{bmatrix} \dot{\theta} \\ \dot{\omega} \\ \dot{i} \end{bmatrix} = \begin{bmatrix} 0 & 1 & 0 \\ 0 & 0 & 4.438 \\ 0 & -12 & -24 \end{bmatrix} \begin{bmatrix} \theta \\ \omega \\ i \end{bmatrix} + \begin{bmatrix} 0 \\ 0 \\ 20 \end{bmatrix} v + \begin{bmatrix} 0 \\ -7.396 \\ 0 \end{bmatrix} T_L,$$

where v is the voltage input and T_L is the load torque. Assume that T_L is a white noise disturbance that is to be rejected.

(a) Assuming that the actuator and sensor are linear, design a controller that achieves 90% disturbance rejection.

(b) Now, assume that the actuator is a saturation of authority $\alpha = 1$. By simulation, determine the level of performance degradation.

(c) Use a-boosting to recover the linear performance. Validate the performance by simulation.

Problem 7.11. Consider the servo model of Problem 7.10. Use conventional LQR to design a state feedback control for the performance index

$$J_{LQR} = \rho \sigma_v^2 + \sigma_\theta^2.$$

For $\rho \in (0,10]$, plot the minimum value of α for which a-boosting is possible.

Problem 7.12. Develop a theory for MIMO boosting. Specifically, consider the LPNI system specified by

$$\dot{x} = Ax + B_1 f(u) + B_2 w$$

$$y = Cx$$

$$y_m = g(y)$$

where $x \in \mathbf{R}^{n_x}$, $u \in \mathbf{R}^{n_u}$, and $y, y_m \in \mathbf{R}^{n_y}$. Let the control be given by

$$u = K y_m,$$

which has been designed with the assumption that all instrumentation is linear. In reality, the instrumentation $f(\cdot)$ and $g(\cdot)$ is nonlinear as in (6.129) and (6.130). The stochastically linearized closed loop system is given by

$$\dot{\hat{x}} = A\hat{x} + B_1 N_a \hat{u} + B_2 w$$

$$\hat{y} = C\hat{x}$$

$$\hat{y}_m = N_s \hat{y}$$

$$\hat{u} = K \hat{y}_m$$

(a) Assuming that the sensor is linear, find conditions for a-boosting, that is, give an expression for K_a such that $N_a K_a = I$.

(b) Assuming that the actuator is linear, find conditions for s-boosting, that is, give an expression for K_s such that $K_s N_s = I$.

(c) Investigate whether a separation principle exists for simultaneous a- and s-boosting?

Problem 7.13. Consider the active suspension system given by

$$\begin{bmatrix} \dot{x}_1 \\ \dot{x}_2 \\ \dot{v}_1 \\ \dot{v}_2 \end{bmatrix} = \begin{bmatrix} 0 & 0 & 1 & 0 \\ 0 & 0 & 0 & 1 \\ -10 & -10 & -2 & 2 \\ 60 & -660 & 12 & -12 \end{bmatrix} \begin{bmatrix} x_1 \\ x_2 \\ v_1 \\ v_2 \end{bmatrix} + \begin{bmatrix} 0 \\ 0 \\ 0.00334 \\ -0.02 \end{bmatrix} u + \begin{bmatrix} 0 \\ 0 \\ 0 \\ 600 \end{bmatrix} d.$$

Assume that the performance output is $z = v_1$. Moreover, assume that all states can be measured.

(a) Design a state feedback control of the form $u = Kx$, where $x = [x_1\ x_2\ v_1\ v_2]$, to achieve 90% disturbance rejection.

(b) Now, assume that the sensor is a MIMO saturation specified by

$$f(\cdot) = [\mathrm{sat}_{0.5}(x_1)\ \mathrm{sat}_{0.5}(x_2)\ \mathrm{sat}_{0.5}(v_1)\ \mathrm{sat}_{0.5}(v_2)].$$

(c) Use the MIMO s-boosting theory derived in the previous problem to recover the performance of part (a).

7.5 Annotated Bibliography

The material of Section 7.1 is based on the following:

[7.1] Y. Eun, C. Gokcek, P.T. Kabamba, and S.M. Meerkov, "Selecting the level of actuator saturation for small performance degradation of linear designs," in *Actuator Saturation Control*, V. Kapila and K.M. Grigiriadis, Eds., Marcel Dekker, Inc., New York, pp. 33–45, 2002

The material of Section 7.2 is based on the following:

[7.2] S. Ching, P.T. Kabamba, and S.M. Meerkov, "Recovery of linear performance in feedback systems with nonlinear instrumentation," *Proceedings of the 2009 American Control Conference*, Vols. 1–9, pp. 2545–2550, 2009

8 Proofs

8.1 Proofs for Chapter 2

Proof of Theorem 2.1. A standard variational argument can be used to determine the optimal $n(t)$. Indeed, for an arbitrary $\delta n(t)$, consider the variation

$$\Delta \varepsilon[n(t), \delta n(t)] = \varepsilon[n(t) + \delta n(t)] - \varepsilon[n(t)]. \tag{8.1}$$

By definition, $n(t)$ minimizes $\varepsilon[n(t)]$ if $\Delta \varepsilon[n(t), \delta n(t)] \geq 0$, for all $\delta n(t)$. Now, from (2.10), it follows that

$$\Delta \varepsilon[n(t), \delta n(t)] = E\{2[\delta n(t) * u(t)][n(t) * u(t) - f[u(t)]] + [\delta n(t) * u(t)]^2\}. \tag{8.2}$$

Thus, $\Delta \varepsilon[n(t), \delta n(t)] \geq 0$, for all $\delta n(t)$ if and only if

$$\delta \varepsilon[n(t), \delta n(t)] = E\{2[\delta n(t) * u(t)][n(t) * u(t) - f[u(t)]]\} = 0, \tag{8.3}$$

for all $\delta n(t)$. Letting $r_{vu}(\tau)$ be the cross-correlation function between $v(t)$ and $u(t)$, and $r_{\hat{v}u}(\tau)$ be the cross-correlation function between $\hat{v}(t)$ and $u(t)$, it is easy to see that $\delta \varepsilon[n(t), \delta n(t)] = 0$, for all $\delta n(t)$ if and only if

$$\int_{-\infty}^{+\infty} [r_{\hat{v}u}(\tau) - r_{vu}(\tau)]\delta n(\tau) d\tau = 0, \tag{8.4}$$

for all $\delta n(t)$. Hence, the optimal $n(t)$ must satisfy

$$r_{\hat{v}u}(\tau) = r_{vu}(\tau). \tag{8.5}$$

Next, since $u(t)$ is a zero-mean Gaussian process, it follows that

$$r_{vu}(\tau) = E\{f'[u(t)]\}r_{uu}(\tau), \tag{8.6}$$

where $r_{uu}(\tau)$ is the autocorrelation of $u(t)$. This result together with

$$r_{\hat{v}u}(\tau) = n(\tau) * r_{uu}(\tau) \tag{8.7}$$

225

implies that the optimal linear system has the impulse response

$$n(t) = E\{f'[u(t)]\}\delta(t), \tag{8.8}$$

where $\delta(t)$ is the unit impulse function.   □

Proof of Theorem 2.3. Under Assumption 2.1, the second term in (2.50), as a function of x, is continuous, and its range covers the range of x. Thus, the existence of a solution is guaranteed by the intermediate value theorem.   □

8.2 Proofs for Chapter 3

Proof of Theorem 3.1. The proof is divided into the following four cases: (a) $0 < |C_0| < \infty$ and $|P_0| < \infty$; (b) $|C_0| = \infty$ and $|P_0| < \infty$; (c) $C_0 = 0$; (d) $|P_0| = \infty$.

Case (a): $0 < |C_0| < \infty$ and $|P_0| < \infty$. For unique e_{ss}^{step}, one of the following takes place: $\lim_{t\to\infty} v(t) = \alpha$, $\lim_{t\to\infty} v(t) = -\alpha$, or $-\alpha < \lim_{t\to\infty} v(t) < \alpha$. Consider first $-\alpha < \lim_{t\to\infty} v(t) < \alpha$. Here, the steady state is identical to that of the linear system, that is,

$$e_{ss}^{step} = \lim_{s\to0} \frac{r_0}{1+P(s)C(s)} = \frac{r_0}{1+P_0C_0}. \tag{8.9}$$

This implies that

$$\lim_{t\to\infty} u(t) = \frac{r_0C_0}{1+P_0C_0}, \tag{8.10}$$

The inequalities $-\alpha < \lim_{t\to\infty} v(t) < \alpha$ hold if and only if

$$|\lim_{t\to\infty} u(t)| < \alpha, \tag{8.11}$$

which takes place if and only if

$$\left|\frac{r_0C_0}{1+P_0C_0}\right| < \alpha. \tag{8.12}$$

This proves statement (i).

In order to show (ii), divide the range of r_0 into four sets: $r_0 > \left|\frac{1}{C_0}+P_0\right|\alpha$, $r_0 < -\left|\frac{1}{C_0}+P_0\right|\alpha$, $r_0 = \left|\frac{1}{C_0}+P_0\right|\alpha$, and $r_0 = -\left|\frac{1}{C_0}+P_0\right|\alpha$. First, we consider the set $r_0 > \left|\frac{1}{C_0}+P_0\right|\alpha$ and show that $e_{ss}^{step} = r_0 - P_0\alpha$ if $C_0 > 0$. For unique e_{ss}^{step}, one of the following takes place: $\lim_{t\to\infty} v(t) = \alpha$ or $\lim_{t\to\infty} v(t) = -\alpha$. Suppose that $\lim_{t\to\infty} v(t) = -\alpha$. Then, $e_{ss}^{step} = r_0 + P_0\alpha$, and

$$\lim_{t\to\infty} u(t) = C_0(r_0+P_0\alpha) \le -\alpha. \tag{8.13}$$

This implies that

$$r_0 \le -\left(\frac{1}{C_0}+P_0\right)\alpha, \tag{8.14}$$

which leads to a contradiction:

$$\left|\frac{1}{C_0} + P_0\right|\alpha < r_0 \le -\left(\frac{1}{C_0} + P_0\right)\alpha. \tag{8.15}$$

Thus, $\lim_{t\to\infty} v(t) = \alpha$, and, consequently, $e_{ss}^{step} = r_0 - P_0\alpha$.

Similarly, one can show that $e_{ss}^{step} = r_0 + P_0\alpha$ if $C_0 < 0$. Combining the two, we obtain $e_{ss}^{step} = r_0 - (\text{sign}C_0)P_0\alpha$.

For the set $r_0 < -\left|\frac{1}{C_0} + P_0\right|\alpha$, similar arguments yield $e_{ss}^{step} = r_0 + (\text{sign}C_0)P_0\alpha$.

Hence, we obtain $e_{ss}^{step} = r_0 - (\text{sign}r_0C_0)P_0\alpha$ for $|r_0| > \left|\frac{1}{C_0} + P_0\right|\alpha$.

Next, we consider $r_0 = \left|\frac{1}{C_0} + P_0\right|\alpha$. Using similar derivations, one can show that $e_{ss}^{step} = r_0 - (\text{sign}C_0)P_0\alpha$ if $(1 + C_0P_0) > 0$. If $(1 + C_0P_0) < 0$, two e_{ss}^{step} exist: $e_{ss}^{step} = r_0 - P_0\alpha$ and $e_{ss}^{step} = r_0 + P_0\alpha$. Since a unique e_{ss}^{step} is assumed, the case of $(1 + C_0P_0) < 0$ is excluded. Similarly, $r_0 = -\left|\frac{1}{C_0} + P_0\right|\alpha$ yields $e_{ss}^{step} = r_0 + (\text{sign}C_0)P_0\alpha$, if e_{ss}^{step} is unique.

Combining the results for all four sets, we conclude that $e_{ss}^{step} = r_0 - (\text{sign}r_0C_0)P_0\alpha$ for $|r_0| \ge \left|\frac{1}{C_0} + P_0\right|\alpha$, which proves (ii).

Case (b): $|C_0| = \infty$ and $|P_0| < \infty$. First, we investigate the situation when $-\alpha < \lim_{t\to\infty} v(t) < \alpha$, that is, when the steady state is identical to that of the linear system and

$$\lim_{t\to\infty} u(t) = \frac{r_0}{P_0}. \tag{8.16}$$

The inequalities $-\alpha < \lim_{t\to\infty} v(t) < \alpha$ hold if and only if

$$|\lim_{t\to\infty} u(t)| < \alpha, \tag{8.17}$$

which takes place if and only if

$$\left|\frac{r_0}{P_0}\right| < \alpha. \tag{8.18}$$

This proves statement (i).

Similarly to the Case (a), we divide the range of r_0 into four sets: $r_0 > |P_0|\alpha$, $r_0 < -|P_0|\alpha$, $r_0 = |P_0|\alpha$, and $r_0 = -|P_0|\alpha$. First we consider the set $r_0 > |P_0|\alpha$ and show that $e_{ss}^{step} = r_0 + P_0\alpha$ if $C_0 = \infty$. For unique e_{ss}^{step}, one of the following takes place: $\lim_{t\to\infty} v(t) = \alpha$ or $\lim_{t\to\infty} v(t) = -\alpha$. Suppose $\lim_{t\to\infty} v(t) = -\alpha$. Then, $e_{ss}^{step} = r_0 + P_0\alpha$, and, since $C_0 = \infty$, $e_{ss}^{step} \le 0$ for $\lim_{t\to\infty} v(t) = -\alpha$ to take place. This implies

$$r_0 + P_0\alpha \le 0, \tag{8.19}$$

which leads to a contradiction:

$$|P_0|\alpha < r_0 \le -P_0\alpha. \tag{8.20}$$

Therefore, $\lim_{t\to\infty} v(t) = \alpha$, and, consequently, $e_{ss}^{step} = r_0 - P_0\alpha$. Similarly, one can show that $e_{ss}^{step} = r_0 + P_0\alpha$ if $C_0 = -\infty$. Combining the two, we obtain $e_{ss}^{step} = r_0 - (\text{sign}C_0)P_0\alpha$.

For $r_0 < -|P_0|\alpha$, similar arguments yield $e_{ss}^{step} = r_0 + (\text{sign}C_0)P_0\alpha$. Hence, we obtain $e_{ss}^{step} = r_0 - (\text{sign}r_0 C_0)P_0\alpha$, if $|r_0| > |P_0|\alpha$.

Finally, consider $r_0 = |P_0|\alpha$. One can show that $e_{ss}^{step} = r_0 - (\text{sign}C_0)P_0\alpha$ if $(1 + C_0 P_0) > 0$. If $(1 + C_0 P_0) < 0$, two e_{ss}^{step} exist: $e_{ss}^{step} = r_0 - P_0\alpha$ and $e_{ss}^{step} = r_0 + P_0\alpha$. Since a unique e_{ss}^{step} is assumed, the case of $(1 + C_0 P_0) < 0$ is excluded. Similarly, $r_0 = -|P_0|\alpha$ yields $e_{ss}^{step} = r_0 + (\text{sign}C_0)P_0\alpha$, if e_{ss}^{step} is unique.

Again, combining the results for all four sets, we conclude that $e_{ss}^{step} = r_0 - (\text{sign}r_0 C_0)P_0\alpha$ for $|r_0| > |P_0|\alpha$, which proves statement (ii).

Case (c): $C_0 = 0$. For this case, we need to show only (i). Since $C(s)$ has a zero at $s = 0$, under the assumption that e_{ss}^{step} exists

$$\lim_{t\to\infty} u(t) = \lim_{s\to 0} e_{ss}^{step} C(s) = 0. \tag{8.21}$$

Therefore, e_{ss}^{step} is identical to that of the linear case. This proves statement (i).

Case (d): $|P_0| = \infty$. For this case, again, we need to show only (i). Since $P(s)$ has a pole at $s = 0$,

$$\lim_{t\to\infty} v(t) = 0 \tag{8.22}$$

must be satisfied for e_{ss}^{step} to exist. Therefore, e_{ss}^{step} is identical to that of the linear case. This proves statement (i). $\square$

The proof of Theorem 3.2 is based on the following two lemmas.

Lemma 8.1. *Assume that with $r_0 = 0$, the equilibrium point $x_p = 0$, $x_c = 0$ of (3.10) is globally asymptotically stable, and this fact can be established by means of a Lyapunov function of the form*

$$V(x) = x^T Q x + \int_0^{Cx} \text{sat}_\alpha(\tau)d\tau > 0, \quad Q \geq 0, \tag{8.23}$$

where $C = [-D_c C_p \ C_c]$ and $x = [x_p^T \ x_c^T]^T$. Let

$$A = \begin{bmatrix} A_p & 0 \\ -B_c C_p & A_c \end{bmatrix}, \ B = \begin{bmatrix} B_p \\ 0 \end{bmatrix}, \ C = [-D_c C_p \ C_c], \tag{8.24}$$

and

$$M = (A + BC)^T (Q + \frac{1}{2}C^T C) + (Q + \frac{1}{2}C^T C)(A + BC), \tag{8.25}$$

where $A_p, B_p, C_p, A_c, B_c, C_c, D_c$ are given in (3.10). Then,

(i) *$A + BC$ is Hurwitz,*
(ii) *$Q + \frac{1}{2}C^T C > 0$,*
(iii) *$M \leq 0$,*
(iv) *$(A + BC, M)$ is observable.*

Proof. Using (8.24), rewrite (3.10) with $r_0 = 0$ as

$$\dot{x} = Ax + B\,\mathrm{sat}_\alpha(u),$$
$$u = Cx. \tag{8.26}$$

If $|Cx| \leq \alpha$, $V(x)$ of (8.23) reduces to

$$V(x) = x^T(Q + \frac{1}{2}C^TC)x, \tag{8.27}$$

the derivative of $V(x)$ along the trajectory of (8.26) becomes

$$\dot{V}(x) = x^T M x, \tag{8.28}$$

and (8.26) can be written as

$$\dot{x} = (A + BC)x. \tag{8.29}$$

Since the closed loop system (8.26) is globally asymptotically stable, $A + BC$ is Hurwitz, which proves (i). Since $V(x) > 0$ and $\dot{V}(x) \leq 0$, matrices $Q + \frac{1}{2}C^TC$ and M are positive definite, and negative semidefinite, respectively. This proves (ii) and (iii). Finally, (i), (ii), and (iii) imply that the pair $(A + BC,\ M)$ is observable, which proves (iv).

Lemma 8.2. *Assume that with $r_0 = 0$, the equilibrium point $x_p = 0$, $x_c = 0$ of (3.10) is globally asymptotically stable, and this fact can be established by means of a Lyapunov function of the form (8.23). Then, for every asymmetric saturation,*

$$\mathrm{sat}_{\alpha\beta}(u) := \mathrm{sat}_\alpha(u + \beta) - \beta$$

$$= \begin{cases} \alpha - \beta, & \text{if } u \geq \alpha - \beta, \\ u, & \text{if } -\alpha - \beta < u < \alpha - \beta, \\ -\alpha - \beta, & \text{if } u \leq -\alpha - \beta, \end{cases} \tag{8.30}$$

with $|\beta| < \alpha$, the closed loop system (3.10) with $r_0 = 0$, that is,

$$\dot{x}_p = A_p x_p + B_p\,\mathrm{sat}_{\alpha\beta}(u),$$
$$\dot{x}_c = A_c x_c + B_c(-y),$$
$$y = C_p x_p, \tag{8.31}$$
$$u = C_c x_c + D_c(-y),$$

is globally asymptotically stable.

Proof. It will be shown that, based on $V(x)$ of (8.23), a Lyapunov function can be found that establishes global asymptotic stability of (8.31) for any $\beta \in (-\alpha, \alpha)$.

Without loss of generality, assume $0 < \beta < \alpha$. Introducing the substitution $\xi = \dfrac{\alpha}{\alpha + \beta}x$ and the notation $\mu = \dfrac{\alpha}{\alpha + \beta}u$, rewrite (8.31) as

$$\dot{\xi} = A\xi + B\psi(\mu),$$
$$\mu = C\xi, \tag{8.32}$$

where A, B, and C are defined in (8.24) and $\psi(\mu) = \dfrac{\alpha}{\alpha+\beta}\mathrm{sat}_{\alpha\beta}\left(\dfrac{\alpha+\beta}{\alpha}\mu\right)$, that is,

$$\psi(\mu) = \begin{cases} \gamma, & \text{if } \mu \geq \gamma, \\ \mu, & \text{if } -\alpha < \mu < \gamma, \\ -\alpha, & \text{if } \mu \leq -\alpha, \end{cases} \tag{8.33}$$

$$\gamma = \frac{\alpha(\alpha-\beta)}{\alpha+\beta} < \alpha.$$

For the closed loop system (8.32), (8.33), select a Lyapunov function candidate as follows:

$$V_1(\xi) = \xi^T Q \xi + \int_0^\mu \psi(\tau)d\tau, \tag{8.34}$$

where Q is the same as in (8.23). The derivative of $V_1(\xi)$ along the trajectories of (8.32), (8.33) is

$$\dot{V}_1(\xi) = \xi^T(A^T Q + QA)\xi + \psi(C\xi)(2B^T Q + CA)\xi + CB[\psi(C\xi)]^2, \tag{8.35}$$

while the derivative of $V(x)$ along the trajectories of (3.10) is

$$\dot{V}(x) = x^T(A^T Q + QA)x + \mathrm{sat}_\alpha(Cx)(2B^T Q + CA)x + CB[\mathrm{sat}_\alpha(Cx)]^2. \tag{8.36}$$

Now we show by contradiction that $\dot{V}_1(\xi) \leq 0$. Assume there exist $\xi^* \neq 0$ such that $\dot{V}_1(\xi^*) > 0$. This ξ^* must satisfy $\psi(C\xi^*) \neq \mathrm{sat}_\alpha(C\xi^*)$, that is, $C\xi^* > \gamma$, otherwise, it would result in $\dot{V}(x) > 0$ at $x = \xi^*$. Define $x^* = \dfrac{\alpha}{\gamma}\xi^*$. Then, $Cx^* > \alpha$, $\psi(C\xi^*) = \gamma$, and $\mathrm{sat}_\alpha(Cx^*) = \alpha$. Substituting x^* in (8.36) yields

$$\begin{aligned}
\dot{V}(x^*) &= x^{*T}(A^T Q + QA)x^* + \mathrm{sat}_\alpha(Cx^*)(2B^T Q + CA)x^* + CB[\mathrm{sat}_\alpha(Cx^*)]^2 \\
&= (\alpha/\gamma)^2\xi^{*T}(A^T Q + QA)\xi^* + \alpha(2B^T Q + CA)\xi^*\alpha/\gamma + CB\alpha^2 \\
&= (\alpha/\gamma)^2[\xi^{*T}(A^T Q + QA)\xi^* + \gamma(2B^T Q + CA)\xi^* + CB\gamma^2] \\
&= (\alpha/\gamma)^2[\xi^{*T}(A^T Q + QA)\xi^* + \psi(C\xi^*)(2B^T Q + CA)\xi^* + CB[\psi(C\xi^*)]^2] \\
&= (\alpha/\gamma)^2\dot{V}_1(\xi^*) > 0,
\end{aligned} \tag{8.37}$$

which contradicts the fact that the global asymptotic stability of system (3.10) is established by (8.23). Therefore, $\dot{V}_1(\xi) \leq 0$ for all $\xi \neq 0$.

Next we show, again by contradiction, that the only solution of (8.32), (8.33) that is contained in $\{\xi : \dot{V}_1(\xi) = 0\}$ is the trivial solution $\xi(t) \equiv 0$. Let $\tilde{\xi}(t)$, $t \geq 0$, be a nontrivial solution of (8.32), (8.33) that satisfies $\dot{V}_1(\tilde{\xi}(t)) \equiv 0$. Assume first $C\tilde{\xi}(t) \leq -\gamma$ for all $t \geq 0$. Since $\mathrm{sat}_\alpha(C\tilde{\xi}(t)) = \psi(C\tilde{\xi}(t))$, system (3.10) and system (8.32), (8.33) are identical; therefore, $\tilde{x}(t) = \tilde{\xi}(t)$ is a nontrivial solution of (3.10) as well. Moreover, as it follows from (8.35) and (8.36), $\dot{V}(\tilde{x}(t)) = \dot{V}_1(\tilde{\xi}(t)) \equiv 0$. This leads to a contradiction. Hence, $\tilde{\xi}(t)$ cannot satisfy $C\tilde{\xi}(t) \leq -\gamma$ for all $t \geq 0$.

Assume now that $C\tilde{\xi}(t) \geq \gamma$ for all $t \geq 0$. Then, from (8.32), (8.33),

$$\tilde{\xi}(t) = e^{At}\tilde{\xi}(0) + \int_0^t e^{A\tau} B\gamma \, d\tau. \tag{8.38}$$

Define $\tilde{x}(t) = \frac{\alpha}{\gamma}\tilde{\xi}(t)$. Then, since $C\tilde{x}(t) > \alpha$ and

$$\tilde{x}(t) = \frac{\alpha}{\gamma}\tilde{\xi}(t) = e^{At}\tilde{x}(0) + \int_0^t e^{A\tau} B\alpha \, d\tau, \tag{8.39}$$

$\tilde{x}(t)$ is a solution of (3.10). Moreover, using the chain of equalities similar to (8.37), we obtain $\dot{V}(\tilde{x}(t)) = (\alpha/\gamma)^2 \, \dot{V}_1(\tilde{\xi}(t)) \equiv 0$. This again leads to a contradiction. Hence, $\tilde{\xi}(t)$ cannot satisfy $C\tilde{\xi}(t) \geq \gamma$ for all $t \geq 0$ either. Therefore, there must exist an interval $(t_1, t_2), t_1 < t_2$, such that $|C\tilde{\xi}(t)| < \gamma$ for all $t \in (t_1, t_2)$. In this interval,

$$\tilde{\xi}(t) = e^{(A+BC)(t-t_1)}\tilde{\xi}(t_1), \quad \forall t \in (t_1, t_2), \tag{8.40}$$

and

$$\dot{V}_1(\tilde{\xi}(t)) = \tilde{\xi}^T(t) M \tilde{\xi}(t) = 0, \quad \forall t \in (t_1, t_2), \tag{8.41}$$

where M is defined in (8.25). Since M is negative semidefinite by Lemma 8.1, it follows that

$$M\tilde{\xi}(t) = M e^{(A+BC)(t-t_1)}\tilde{\xi}(t_1) = 0, \quad \forall t \in (t_1, t_2). \tag{8.42}$$

This, however, contradicts the observability of $(A + BC, M)$, which must take place according to Lemma 8.1. Thus, $\tilde{\xi}(t) \equiv 0$ is the only solution of (8.32), (8.33) contained in $\{\xi : \dot{V}_1(\xi) = 0\}$.

Finally, it can be shown, again by contradiction, that $V_1(\xi) > 0$ for all $\xi \neq 0$ and that $V_1(\xi) \to \infty$ as $|\xi| \to \infty$. Therefore, the system (8.32) and (8.33), and, hence, (8.31) is globally asymptotically stable.

Proof of Theorem 3.2. The proof is provided for the following three cases separately: (a) $|C_0| < \infty$ and $|P_0| < \infty$; (b) $|C_0| = \infty$ and $|P_0| < \infty$; (c) $|P_0| = \infty$.

Case (a): $|C_0| < \infty$ and $|P_0| < \infty$. Define

$$u^* = \frac{C_0}{1 + C_0 P_0} r_0, \quad y^* = \frac{C_0 P_0}{1 + C_0 P_0} r_0, \tag{8.43}$$

$$x_p^* = -A_p^{-1} B_p u^*, \quad x_c^* = -A_c^{-1} B_c (r_0 - y^*).$$

Using

$$\hat{x}_p = x_p - x_p^*, \quad \hat{x}_c = x_c - x_c^*, \quad \hat{y} = y - y^*, \quad \hat{u} = u - u^*, \tag{8.44}$$

rewrite (3.10) as

$$\begin{aligned}
\dot{\hat{x}}_p &= A_p \hat{x}_p + B_p \, \text{sat}_{\alpha u^*}(\hat{u}), \\
\dot{\hat{x}}_c &= A_c \hat{x}_c + B_c(-\hat{y}), \\
\hat{y} &= C_p \hat{x}_p, \\
\hat{u} &= C_c \hat{x}_c + D_c(-\hat{y}),
\end{aligned} \tag{8.45}$$

where $\text{sat}_{\alpha u^*}(\hat{u}) = \text{sat}_\alpha(\hat{u} + u^*) - u^*$.

Note that the condition $r_0 \in$ TD implies $|u^*| < \alpha$. Therefore, Lemma 8.2 ensures the global asymptotic stability of (8.45). Hence, the steady state exists and $y(t)$ converges to y^*, that is, a unique e_{ss}^{step} exists.

Case (b): $|C_0| = \infty$ implies that A_c has an eigenvalue 0. Choose $x_c^* \neq 0$ to satisfy

$$A_c x_c^* = 0, \quad C_c x_c^* = \frac{r_0}{P_0}, \tag{8.46}$$

and let

$$u^* = \frac{r_0}{P_0}, \quad y^* = r_0, \quad x_p^* = -A_p^{-1} B_p u^*. \tag{8.47}$$

Proceeding similarly to (a), it can be shown that if $r_0 \in$ TD, then a unique e_{ss}^{step} exists.

Case (c): $|P_0| = \infty$ implies A_p has an eigenvalue 0. Choose $x_p^* \neq 0$ to satisfy

$$A_p x_p^* = 0, \quad C_p x_p^* = r_0, \tag{8.48}$$

and let $u^* = 0$, $y^* = r_0$, $x_c^* = 0$. Similarly to (a), it can be shown that a unique e_{ss}^{step} exists. This completes the proof. $\qquad\qquad\square$

Proof of Theorem 3.3. *Part (i):* Consider the system shown in Figure 8.1(a) with ramp input. Equivalently, it can be represented as shown in Figure 8.1(b), where the input is step. We refer to these systems as system (a) and (b), respectively. Since the input to system (b) is a step function, e_{ss}^{ramp} of system (a) can be analyzed using Theorem 3.1 applied to system (b). Indeed, introducing

$$\widehat{P}(s) = sP(s), \quad \widehat{C}(s) = \frac{1}{s}C(s), \tag{8.49}$$

and noting that

$$\widehat{P}_0 = P_1, \quad \widehat{C}_0 = \infty, \tag{8.50}$$

condition (3.29) for system (a) is equivalent to $|r_1| < \left| \dfrac{1}{\widehat{C}_0} + \widehat{P}_0 \right| \alpha$ for system (b). Then, it follows from the proof of Theorem 3.1 that $|\lim_{t \to \infty} u(t)| < \alpha$. Therefore, e_{ss}^{ramp} must be identical to that of the system with linear actuator, that is,

$$e_{ss}^{ramp} = \frac{r_1}{\lim_{t \to \infty} sP(s)C(s)}. \tag{8.51}$$

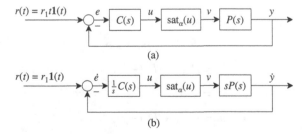

Figure 8.1. Block diagram of a system with ramp input and its equivalent representation.

Since $P_1 = \lim_{t \to \infty} sP(s) \neq 0$, it follows that

$$\lim_{t \to \infty} sP(s)C(s) = \lim_{t \to \infty} sP(s) \lim_{t \to \infty} C(s) = P_1 C_0, \tag{8.52}$$

and

$$e_{ss}^{\mathrm{ramp}} = \frac{r_1}{P_1 C_0}, \tag{8.53}$$

which proves statement (i).

Part (ii) is proved similarly. $\qquad\square$

Proof of Theorem 3.4. *Part (i):* Due to the stability assumption, $|S(j\omega)|$ is continuous and bounded. Given $\epsilon > 0$, choose ω_0 such that

$$\left| |S(j0)|^2 - |S(j\omega)|^2 \right| < \frac{\epsilon}{2}, \quad \forall \omega < \omega_0 \tag{8.54}$$

is satisfied. Using (8.54) and the fact that $\|F_\Omega(s)\|_2 = 1$, the following holds:

$$
\begin{aligned}
\left| RS(\Omega)^2 - |S(j0)|^2 \right| &= \left| \frac{1}{\pi} \int_0^\infty |F_\Omega(j\omega)|^2 \left(|S(j\omega)|^2 - |S(j0)|^2 \right) d\omega \right| \\
&\leq \frac{1}{\pi} \int_0^\infty |F_\Omega(j\omega))|^2 \left| |S(j\omega)|^2 - |S(j0)|^2 \right| d\omega \\
&= \frac{\epsilon}{2} \frac{1}{\pi} \int_0^{\omega_0} |F_\Omega(j\omega)|^2 d\omega + M_r^2 \frac{1}{\pi} \int_{\omega_0}^\infty |F_\Omega(j\omega)|^2 d\omega \\
&< \frac{\epsilon}{2} + M_r^2 \frac{1}{\pi} \int_{\omega_0}^\infty |F_\Omega(j\omega)|^2 d\omega.
\end{aligned}
\tag{8.55}
$$

The term $M_r^2 \frac{1}{\pi} \int_{\omega_0}^\infty |F_\Omega(j\omega)|^2 d\omega$ can be made smaller than $\frac{\epsilon}{2}$ if Ω is sufficiently small. Hence, we obtain

$$\left| RS(\Omega)^2 - |S(j0)|^2 \right| < \epsilon, \tag{8.56}$$

for Ω sufficiently small. This proves statement (i).

Part (ii) is proved analogously.

Part (iii) follows directly from (3.50). $\qquad\square$

Proof of Theorem 3.5. *Part (i):* It is sufficient to show that equation (3.75) is satisfied with $N = 1$. By assumption, $\left\| \frac{F_\Omega(s)C(s)}{1+P(s)C(s)} \right\|_2$ is finite for any Ω. As σ_r tends to zero, due to the property of $\mathrm{erf}(\cdot)$, $N = 1$ solves equation (3.75).

Part (ii): We show that as Ω tends to infinity, the quantity $\left\| \frac{F_\Omega(s)C(s)}{1+NP(s)C(s)} \right\|_2$ in (3.75) converges to C_∞ regardless of N, where $C_\infty = \lim_{s \in \mathbf{R},\, s \to \infty} |C(s)|$. Indeed, given $\varepsilon > 0$, choose ω_0 such that

$$\left| C_\infty^2 - \left| \frac{C(j\omega)}{1+NP(j\omega)C(j\omega)} \right|^2 \right| < \frac{\varepsilon}{2}, \quad \forall \omega > \omega_0. \tag{8.57}$$

Such ω_0 exists since $P(s)$ is strictly proper and $C(s)$ is proper. Then, using (8.57) and the fact that $||F_\Omega(s)||_2 = 1$, the following holds:

$$\left| C_\infty^2 - \left\| \frac{F_\Omega(s)C(s)}{1+NP(s)C(s)} \right\|_2 \right| = \left| \frac{1}{\pi} \int_0^\infty |F_\Omega|^2 \left(C_\infty^2 - \left| \frac{C}{1+NPC} \right|^2 \right) d\omega \right|$$

$$\leq \frac{1}{\pi} \int_0^{\omega_0} |F_\Omega|^2 \left| C_\infty^2 - \left| \frac{C}{1+NPC} \right|^2 \right| d\omega + \frac{1}{\pi} \int_{\omega_0}^\infty |F_\Omega|^2 \frac{\varepsilon}{2} d\omega$$

$$< \frac{1}{\pi} \frac{3}{\Omega} \int_0^{\omega_0} \left| C_\infty^2 - \left| \frac{C}{1+NPC} \right|^2 \right| d\omega + \frac{\varepsilon}{2}. \tag{8.58}$$

As Ω tends to infinity, the term $\frac{1}{\pi} \frac{3}{\Omega} \int_0^{\omega_0} \left| C_\infty^2 - \left| \frac{C}{1+NPC} \right|^2 \right| d\omega$ can be made smaller than $\varepsilon/2$, and, thus,

$$\left| C_\infty^2 - \left\| \frac{F_\Omega(s)C(s)}{1+NP(s)C(s)} \right\|_2 \right| < \varepsilon. \tag{8.59}$$

Hence, the solution of (3.75) as $\Omega \to \infty$ is given by

$$N_1 = \mathrm{erf}\left(\frac{\alpha}{\sqrt{2}\sigma_r C_\infty} \right). \tag{8.60}$$

Similarly, it can be shown that $\lim_{\Omega \to \infty} ||F_\Omega/(1+N_1 PC)||_2 = 1$. Thus,

$$\lim_{\Omega \to \infty} SRS(\Omega, \sigma_r) = \lim_{\Omega \to \infty} \lim_{N \to N_1} \left\| \frac{F_\Omega}{1+NPC} \right\|_2 = 1, \tag{8.61}$$

which proves (ii).

Part (iii) is proved similarly.   □

8.3 Proofs for Chapter 4

Proof of Theorem 4.1. A straightforward linear analysis applied to (4.23) yields $\mathrm{diag}\{\sigma_{\hat{u}}^2, \sigma_{\hat{y}}^2\} = \mathrm{diag}\{\tilde{K}\tilde{P}\tilde{K}^T\}$, where $\tilde{P}$ is the solution of the Lyapunov equation (4.25). Using this result in (4.20) and (4.21) gives equation (4.26). Finally, equation (4.24) follows from $\sigma_{\hat{z}}^2 = \mathrm{tr}\{\tilde{C}\tilde{R}\tilde{C}^T\}$.   □

Proof of Theorem 4.2. We need the following auxiliary result for the proof. Assume $R \in \mathbf{R}^{n \times n}$, $R = R^T > 0$ and $\tilde{A} \in \mathbf{R}^{n \times n}$ is Hurwitz, then the solutions P and Q of

$$\tilde{A}P + P\tilde{A}^T + R = 0, \tag{8.62}$$

$$\tilde{A}Q + Q\tilde{A}^T + R \leq 0, \tag{8.63}$$

satisfy $P = P^T > 0$, $Q = Q^T > 0$ and

$$P \leq Q. \tag{8.64}$$

This result can be easily verified by noting that (8.62) and (8.63) yield

$$\tilde{A}(Q-P)+(Q-P)\tilde{A}^T \leq 0 \tag{8.65}$$

and this implies

$$Q-P \geq 0 \tag{8.66}$$

by the Lyapunov theorem.

Now, assume that $(A+BNK)$ is Hurwitz. Then there exists a unique symmetric positive definite solution of

$$(A+BNK)P+P(A+BNK)^T+BB^T=0, \tag{8.67}$$

and the variance at the output is less than or equal to γ^2 if

$$CPC^T \leq \gamma^2. \tag{8.68}$$

Let Q solve

$$(A+BNK)Q+Q(A+BNK)^T+BB^T \leq 0. \tag{8.69}$$

Then, by the above result, $P \leq Q$. So,

$$CPC^T \leq CQC^T, \tag{8.70}$$

which implies that (8.68) is satisfied whenever

$$CQC^T \leq \gamma^2 \tag{8.71}$$

is satisfied.   □

Proof of Theorem 4.3. We first show that

$$KCQC^TK^T - \frac{2\alpha^2}{\pi} < 0, \tag{8.72}$$

and

$$(A+BNK)Q+Q(A+BNK)^T+BB^T \leq 0. \tag{8.73}$$

hold if and only if

$$N=f(N) \tag{8.74}$$

has a solution $N > 0$ for

$$f(N) := \mathrm{erf}\left(\frac{\alpha N}{\sqrt{2KQK^T}}\right). \tag{8.75}$$

Indeed, function $f(N)$ is monotonically increasing on $[0,\infty)$, concave on $[0,\infty)$, bounded from above by 1 on $[0,\infty)$, and is zero at the origin. Therefore, (8.74) has a solution $N > 0$ if and only if $\frac{d^2}{dN^2}f(N)\big|_{N=0} > 1$. This is true if and only if (8.72), (8.73) are satisfied. Now, (8.74), (8.75) are exactly (4.49) subject to (4.50), (4.51), which, when solved, give the exact value for N. Thus, the LMIs (4.53) and (4.54) follow directly from this result with $Y = K\bar{P}$ and (4.53) is the Schur complement of (8.72). Inequality (4.55) follows from Theorem 4.2 and ensures the desired performance.

$\square$

8.4 Proofs for Chapter 5

Proof of Lemma 5.1. *Part (i):* To prove continuity, note that (5.34) can be rewritten as

$$F(K, K_e(K)) = 0, \tag{8.76}$$

where

$$F(x, y) = y - x\mathrm{erf}\left(\frac{\alpha}{\sqrt{2}x\left\|\frac{F_\Omega(s)C(s)}{1+yP(s)C(s)}\right\|_2}\right) \tag{8.77}$$

is an analytic function. Hence, $K_e(K)$ is a root of an analytic equation that depends on a parameter K. As it is well known, the roots of such an equation are continuous with respect to the parameter, and the result follows.

The proof of strict monotonicity is by contradiction. Indeed, let us assume that there exists K_1, K_2 such that $K_2 > K_1$ and $K_e(K_1) = K_e(K_2) = K_e^*$. This implies that

$$K_2\mathrm{erf}\left(\frac{\alpha}{\sqrt{2}K_2\sigma^*}\right) = K_1\mathrm{erf}\left(\frac{\alpha}{\sqrt{2}K_1\sigma^*}\right), \tag{8.78}$$

where

$$\sigma^* = \left\|\frac{F_\Omega(s)C(s)}{1+K_e^*C(s)P(s)}\right\|_2. \tag{8.79}$$

However, it is straightforward to verify that the function

$$f(x) = x\mathrm{erf}\left(\frac{c}{x}\right) \tag{8.80}$$

is strictly monotonic for all $c > 0$. Hence (8.78) implies $K_1 = K_2$, which contradicts the assumption that $K_2 > K_1$. Therefore, $K_e(K)$ must be a strictly monotonic function.

Part (ii): Note that (5.34) can be expressed equivalently as

$$K_e(K) = K\mathrm{erf}\left(\frac{\alpha}{\sqrt{2}K\phi(K)}\right), \tag{8.81}$$

where

$$\phi(K) = \left\|\frac{F_\Omega(s)C(s)}{1+K_e(K)P(s)C(s)}\right\|_2. \tag{8.82}$$

By expanding erf(·) in (8.81), in Taylor series, one obtains

$$K_e(K) = \frac{\sqrt{2}K}{\sqrt{\pi}} \left(\frac{\alpha}{K\phi(K)} - \frac{1}{3} \left(\frac{\alpha}{K\phi(K)} \right)^3 + \dots \right). \tag{8.83}$$

It follows from continuity and strict monotonicity that the limit of $K_e(K)$ as $K \to \infty$ either exists or is infinity. The remainder of the proof is by contradiction. Namely, assume that β_1 and β_2 are any two distinct positive solutions of (5.38), while $K_e(K)$ is unique for all K. Taking the limit of (8.82) and (8.83) results in

$$\lim_{K \to \infty} K_e(K) = \lim_{K \to \infty} \left(\frac{\alpha\sqrt{2/\pi}}{\phi(K)} - \frac{\sqrt{2/\pi}}{3} \left(\frac{\alpha^3}{K^2\phi^3(K)} \right) + \dots \right) \tag{8.84}$$

and

$$\lim_{K \to \infty} \phi(K) = \left\| \frac{F_\Omega(s)C(s)}{1 + \left(\lim_{K \to \infty} K_e(K) \right) P(s)C(s)} \right\|_2. \tag{8.85}$$

These constitute equations for the limits of $K_e(K)$ and $\phi(K)$ when K tends to infinity. Two possible solutions of (8.84) and (8.85) exist, given by:

$$\lim_{K \to \infty} K_e(K) = \frac{\alpha\sqrt{2/\pi}}{\beta_i}.$$

$$\lim_{K \to \infty} \phi(K) = \beta_i, i = 1, 2.$$

Clearly, this contradicts the assumption of uniqueness of solution of $K_e(K)$ and hence, (5.38) cannot admit multiple positive solutions.   □

Proof of Theorem 5.1. *Part (i):*

Necessity: Assume (5.39). Take the limit of (8.82) as $K \to \infty$, using the assumption that $T_\gamma(s)$ is stable for all $\gamma > 0$, to obtain

$$\lim_{K \to \infty} \phi(K) = \left\| \frac{F_\Omega(s)C(s)}{1 + \left(\frac{\alpha\sqrt{2/\pi}}{\beta} \right) P(s)C(s)} \right\|_2 \triangleq \phi, \tag{8.86}$$

where $\phi > 0$. By taking the limit of (8.83) it follows from (8.86) that

$$\lim_{K \to \infty} K_e(K) = \frac{\alpha\sqrt{2/\pi}}{\phi}. \tag{8.87}$$

Hence, from (5.39) we obtain

$$\phi = \beta. \tag{8.88}$$

Substituting (8.88) into (8.86) yields (5.38). The uniqueness of solution of (5.38) is guaranteed by Lemma 5.1.

Sufficiency: Assume (5.38) admits a unique solution $\beta > 0$. Recall that (8.84) and (8.85) are equations for the limits of $K_e(K)$ and $\phi(K)$ when K tends to infinity, and are satisfied by

$$\lim_{K \to \infty} K_e(K) = \frac{\alpha\sqrt{2/\pi}}{\beta},\tag{8.89}$$

$$\lim_{K \to \infty} \phi(K) = \beta.\tag{8.90}$$

The right-hand sides of (8.89) and (8.90) are the unique solutions of (8.84) and (8.85). Indeed, suppose that

$$\lim_{K \to \infty} K_e(K) = \frac{\alpha\sqrt{2/\pi}}{\beta_l},$$

$$\lim_{K \to \infty} \phi(K) = \phi_l$$

also satisfy (8.84) and (8.85). It follows from necessity that β_l must be a solution of (5.38), and thus by Lemma 5.1, $\beta_l = \beta$. Then, (8.86) implies $\phi_l = \beta$. Hence, (8.89) and (8.90) are the only solutions of (8.84) and (8.85). The proof concludes by noting that (8.89) yields (5.39).

Part (ii):

Sufficiency: Recall that the limit of $K_e(K)$ either exists or is infinite. Hence, by part (i), if $\beta = 0$ is the only real solution of (5.38), then the limit of $K_e(K)$ must be infinity.

Necessity: Assume (5.40). Then it follows from part (i) that (5.38) cannot admit a unique positive solution. By Lemma 5.1, (5.38) cannot admit multiple positive solutions, and hence it follows that $\beta = 0$ must be the only real solution of (5.38). □

To prove Theorem 5.2 we need the following lemma.

Lemma 8.3. *Let $T_\gamma(s)$ be asymptotically stable only for $\gamma \in [0, \Gamma)$, $\Gamma < \infty$, so that (5.37) holds, and (5.34) admits a unique solution for all $K > 0$. Then*

$$K_e(K) < \Gamma \quad \forall K > 0.\tag{8.91}$$

Proof. The proof is by contradiction. Assume that there exists $K^* > 0$ such that $K_e(K^*) \geq \Gamma$. Then it follows from (5.34) that

$$K_e(K^*) = K^* \mathrm{erf}\left(\frac{\alpha}{\sqrt{2}K^* \left\| \frac{F_\Omega(s)C(s)}{1+K_e(K^*)P(s)C(s)} \right\|_2}\right).\tag{8.92}$$

This, however, is a contradiction because the left-hand side of (8.92) is equal to a positive number, while the right-hand side is 0. □

Proof of Theorem 5.2. By Lemma 8.3, the limit of $K_e(K)$ as K tends to infinity must exist and satisfy

$$\lim_{K \to \infty} K_e(K) \leq \Gamma.$$

The remainder of the proof follows is analogous to that of Theorem 5.1. □

Proof of Theorem 5.3. *Part (i):* We will prove that for any $K_e^d > 0$ there exists K^d such that $K_e(K^d) = K_e^d$. Consider $F(K^d, K_e^d)$, where $F(\cdot, \cdot)$ is defined in (8.77). Clearly,

$$F\left(0, K_e^d\right) = K_e^d > 0. \tag{8.93}$$

Moreover, Taylor series expansion (8.83) of erf$(\cdot)$ in (8.77) yields,

$$\lim_{K^d \to \infty} F\left(K^d, K_e^d\right) = K_e^d - \frac{\alpha\sqrt{2/\pi}}{\left\|\frac{F_\Omega(s)C(s)}{1 + K_e^d P(s)C(s)}\right\|_2}. \tag{8.94}$$

Note that every positive solution β of (5.38) is related to a finite root K_e^d of the right-hand side of (8.94) through

$$K_e^d = \frac{\alpha\sqrt{2/\pi}}{\beta},$$

and vice versa. It follows from (8.94) that

$$\lim_{K^d \to \infty} F\left(K^d, 0\right) = -\frac{\alpha\sqrt{2/\pi}}{\|F_\Omega(s)\,C(s)\|_2} < 0. \tag{8.95}$$

Since (5.38) admits no solution $\beta > 0$, the right-hand side of (8.94) has no finite roots. Moreover, the right-hand side of (8.94) is a continuous function of K_e^d. Hence, (8.95) implies that

$$\lim_{K^d \to \infty} F\left(K^d, K_e^d\right) < 0. \tag{8.96}$$

Recall that $F(K^d, K_e^d)$ is continuous and monotonically decreasing in K^d. It thus follows from (8.93) and (8.96) that for any $K_e^d > 0$, $F(K^d, K_e^d)$ changes sign exactly once as K^d goes from 0 to ∞. Hence, there exists a unique K^d such that $F(K^d, K_e^d) = 0$. Thus, $K_e(K^d) = K_e^d$, and the result follows immediately.

Part (ii): Note that for $K = 0$, (5.34) admits a unique solution $K_e(0) = 0$. Hence, by continuity, (5.34) defines a unique $K_e(K)$ when K is small enough. The S-origination point occurs when (5.34) starts admitting multiple solutions. □

Proof of Theorem 5.4. *Part (i):* By the continuity of roots of (8.77), all solutions $K_e(K)$ of (5.34) must be continuous in K. The remainder of the proof is analagous to that of sufficiency in Theorem 5.1. Namely, it follows that for each β_i,

$$\lim_{K \to \infty} K_e(K) = \frac{\alpha\sqrt{2/\pi}}{\beta_i} \tag{8.97}$$

and

$$\lim_{K \to \infty} \phi(K) = \beta_i \tag{8.98}$$

are valid solutions for the limiting values of $K_e(K)$ and $\phi(K)$ (i.e., (8.84) and (8.85)). Clearly, (8.97) yields (5.51).

Part (ii): It follows from (5.34) that for any K, stochastic linearization must yield an odd number of solutions. Hence, if (5.38) yields an even number of (finite) solutions for the limit of $K_e(K)$, an additional solution must exist, corresponding to the solution $\beta_0 = 0$.

Part (iii): Let κ_1 denote the limiting gain (given by Part (i)) corresponding to the largest simple root of (5.38), and define K_e^t such that

$$\kappa_1 < K_e^t < \kappa_2,$$

where κ_2 denotes the next largest (if any) limiting gain. Clearly, from (8.93),

$$F\left(0, K_e^t\right) > 0, \tag{8.99}$$

and since $K_e^d = \kappa_1$ is the smallest simple root of the right-hand side of (8.94), it follows from continuity and (8.95) that

$$\lim_{K^d \to \infty} F\left(K^d, K_e^t\right) > 0. \tag{8.100}$$

Recalling that $F(K^d, K_e^t)$ is continuous and monotonically decreasing in K^d, it follows from (8.99) and (8.100) that

$$F\left(K^d, K_e^t\right) \neq 0 \quad \forall K^d > 0. \tag{8.101}$$

Thus, $K_e(K^d) = K^t$ cannot be satisfied for any $K^d > 0$, and hence $K_e(K)$ cannot lie in the interval (κ_1, κ_2). The result follows directly. $\qquad\square$

Proof of Theorem 5.5. The proof is by construction. Let $C(s)$ be given by the inverse of the plant, that is,

$$C(s) = \frac{1}{P(s)}.$$

Then (5.34) reduces to

$$K_e = K \mathrm{erf}\left(\frac{\alpha}{\sqrt{2}K \, \|F_\Omega(s)/P(s)\|_2}(1 + K_e)\right). \tag{8.102}$$

For any fixed $K > 0$, the right-hand side of (8.102) is a strictly concave function of K_e, starting from a positive value

$$K\mathrm{erf}\left(\frac{\alpha}{\sqrt{2}\,\|F_\Omega(s)/P(s)\|_2}\right) > 0$$

at $K_e = 0$ and bounded by K for all $K_e > 0$. Moreover, the left-hand side of (8.102) is K_e. Hence, there exists a unique $K_e > 0$ that satisfies (8.102). $\qquad\square$

8.5 Proofs for Chapter 6

Proof of Theorem 6.1. Let Ω be the set $\Omega = \{K : A + B_2NK \text{ is Hurwitz}\}$. Then, for any $K \in \Omega$,

$$\sigma_{\hat{z}}^2 = \text{tr}\{C_1 R C_1^T\} + \rho K R K^T, \tag{8.103}$$

where (N, R) satisfies

$$(A + B_2NK)R + R(A + B_2NK)^T + B_1 B_1^T = 0, \tag{8.104}$$

$$KRK^T - \frac{1}{2\left[\text{erf}^{-1}(N)\right]^2} = 0. \tag{8.105}$$

Hence, the problem at hand is equivalent to

$$\underset{K \in \Omega}{\text{minimize}}\,\text{tr}\{C_1 R C_1^T\} + \rho K R K^T,$$

$$\text{subject to}\,(A + B_2NK)R + R(A + B_2NK)^T + B_1 B_1^T = 0, \tag{8.106}$$

$$KRK^T - \frac{1}{2\left[\text{erf}^{-1}(N)\right]^2} = 0.$$

We use the Lagrange multiplier method to solve this constrained optimization problem.

First, we check the regularity of the constraints. Let (K, N, R) be such that it satisfies the constraints (8.104) and (8.105), and for an arbitrary symmetric matrix Q and real number λ, define the function

$$\Phi(K, N, R) = \text{tr}\{[(A + B_2NK)R + R(A + B_2NK)^T + B_1 B_1^T]Q\}$$

$$+ \lambda \left(KRK^T - \frac{1}{2\left[\text{erf}^{-1}(N)\right]^2} \right). \tag{8.107}$$

Then we have

$$\frac{\partial \Phi}{\partial K} = 0 \Rightarrow NB_2^T QR + \lambda KR = 0, \tag{8.108}$$

$$\frac{\partial \Phi}{\partial N} = 0 \Rightarrow KRQB_2 + \lambda \frac{\sqrt{\pi}}{4} \frac{\exp\left(\left[\text{erf}^{-1}(N)\right]^2\right)}{\left[\text{erf}^{-1}(N)\right]^3} = 0, \tag{8.109}$$

$$\frac{\partial \Phi}{\partial R} = 0 \Rightarrow (A + B_2NK)^T Q + Q(A + B_2NK) + \lambda K^T K = 0. \tag{8.110}$$

Multiplying (8.108) from right by K^T and using (8.105), (8.109) we obtain

$$\lambda \left[\frac{1}{2\left[\text{erf}^{-1}(N)\right]^2} - \frac{\sqrt{\pi}}{4} \frac{N}{\left[\text{erf}^{-1}(N)\right]^3} \exp\left(\left[\text{erf}^{-1}(N)\right]^2\right) \right] = 0. \tag{8.111}$$

Here, since the expression that multiplies λ is nonzero for all $N \in (0,1)$ and as $N \to 0^+$, $N \to 1^-$, we conclude that $\lambda = 0$. Then, with $\lambda = 0$, it follows from (8.110) that $Q = 0$. Hence, the regularity conditions are satisfied.

Next, we form the Lagrangian

$$
\begin{aligned}
\Psi(K,N,R,Q,\lambda) = {}& \mathrm{tr}\{C_1 R C_1^T\} + \rho\, KRK^T \\
& + \mathrm{tr}\{[(A + B_2 NK)R + R(A + B_2 NK)^T + B_1 B_1^T]Q\} \\
& + \lambda\left(KRK^T - \frac{1}{2\left[\mathrm{erf}^{-1}(N)\right]^2}\right),
\end{aligned}
\tag{8.112}
$$

where Q and λ are the Lagrange multipliers. Differentiating Ψ with respect to K, N, R, Q, and λ, and equating the results to zero, we obtain

$$
[(\rho + \lambda)K + NB_2^T Q]R = 0,
\tag{8.113}
$$

$$
KRQB_2 + \lambda\frac{\sqrt{\pi}}{4} \frac{\exp\left(\left[\mathrm{erf}^{-1}(N)\right]^2\right)}{\left[\mathrm{erf}^{-1}(N)\right]^3} = 0,
\tag{8.114}
$$

$$
(A + B_2 NK)^T Q + Q(A + B_2 NK) + (\rho + \lambda)K^T K + C_1^T C_1 = 0,
\tag{8.115}
$$

$$
(A + B_2 NK)R + R(A + B_2 NK)^T + B_1 B_1^T = 0,
\tag{8.116}
$$

$$
KRK^T - \frac{1}{2\left[\mathrm{erf}^{-1}(N)\right]^2} = 0.
\tag{8.117}
$$

Now, it follows from (8.113) that

$$
K = -\frac{N}{\rho + \lambda} B_2^T Q + K_n,
\tag{8.118}
$$

for an arbitrary K_n in the left null space of R that makes $A + B_2 NK$ Hurwitz. However, since $K_n R = 0$, the last term in (8.118) does not affect the value of the performance measure. Thus, we let

$$
K = -\frac{N}{\rho + \lambda} B_2^T Q.
\tag{8.119}
$$

Substituting this expression into (8.115), (8.116), and (8.117) immediately gives (6.10), (6.11), and (6.12), respectively. Multiplying (8.113) from right by K^T and using (8.114) together with (8.117) yields (6.13). Moreover, assuming this K is in Ω and substituting it into (8.103) yields (6.8).

Finally, we show that the system of equations (6.10)–(6.13) has a unique solution for (N,R,Q,λ) such that $K \in \Omega$. Clearly, for $N \in (0,1)$, equation (6.13) defines λ as a continuous function of N and note that $\lambda = \lambda(N) > 0$. Thus, substituting $\lambda = \lambda(N)$ into (6.10), we obtain the Riccati equation

$$
A^T Q + QA - \frac{N^2}{\rho + \lambda(N)} QB_2 B_2^T Q + C_1^T C_1 = 0.
\tag{8.120}
$$

Since (A, B_2) is stabilizable and (C_1, A) is detectable for any $N \in (0,1)$, this Riccati equation has a unique positive semidefinite solution for $Q = Q(N)$ such that $A + B_2 NK(N)$ is Hurwitz, where

$$K(N) = -\frac{N}{\rho + \lambda(N)} B_2^T Q(N). \tag{8.121}$$

Similarly, substituting $\lambda = \lambda(N)$ and $Q = Q(N)$ into (6.11), we obtain the Lyapunov equation

$$[A - \frac{N^2}{\rho + \lambda(N)} B_2 B_2^T Q(N)]R + R[A - \frac{N^2}{\rho + \lambda(N)} B_2 B_2^T Q(N)]^T \\ + B_1 B_1^T = 0. \tag{8.122}$$

Since $A + B_2 NK(N)$ is Hurwitz, for any $N \in (0,1)$, this Lyapunov equation has also a unique positive semidefinite solution for $R = R(N)$. In addition, as $N \to 1^-$, λ approaches zero, Q and R remain finite, and $A + B_2 NK$ remains Hurwitz. Hence, in order to show that the system of equations (6.10)–(6.13) has a unique solution, it is sufficient to show that the equation

$$\left[\frac{N^2}{\rho + \lambda(N)}\right]^2 B_2^T Q(N)R(N)Q(N)B_2 = \frac{N^2 \alpha^2}{2\left[\text{erf}^{-1}(N)\right]^2} \tag{8.123}$$

has a unique solution for N. The right-hand side of this equation is a strictly decreasing function of N for all $N \in (0,1)$, and moreover, it assumes the values $2/\pi$ and 0 as $N \to 0^+$ and $N \to 1^-$, respectively. Now, we show that the left-hand side of this equation is an increasing function of N and approaches 0 as $N \to 0^+$. For $N \in (0,1)$, define

$$\tau(N) := \frac{\rho + \lambda(N)}{N^2}. \tag{8.124}$$

Note that $\tau(N)$ is a continuous function of N and it decreases monotonically from ∞ to ρ as N increases from 0 to 1. Moreover, define the function $f : (0,1) \to \mathbf{R}^+$ as

$$f[\tau(N)] := \frac{1}{\tau^2(N)} B_2^T Q[\tau(N)]R[\tau(N)]Q[\tau(N)]B_2, \tag{8.125}$$

which is nothing but the left-hand side of (8.123). Note that $f(N)$ is also a continuous function of N. Differentiating $f[\tau(N)]$ with respect to N, we obtain

$$\frac{d}{dN}f[\tau(N)] = \frac{d}{d\tau}f(\tau)\frac{d}{dN}\tau(N). \tag{8.126}$$

Since $\tau(N)$ is a decreasing function of N, it follows that $\tau'(N) < 0$ for all $N \in (0,1)$. Thus, to show that $f(N)$ is an increasing function of N, it remains to show that $f'(\tau) \leq 0$ for all $\tau \in (\rho, \infty)$. Substituting (8.124) into (8.120) and (8.122), respectively, it follows that $Q(\tau)$ and $R(\tau)$ satisfy

$$[A - \frac{1}{\tau}B_2 B_2^T Q(\tau)]^T Q(\tau) + Q(\tau)[A - \frac{1}{\tau}B_2 B_2^T Q(\tau)] \\ + \frac{1}{\tau}Q(\tau)B_2 B_2^T Q(\tau) + C_1^T C_1 = 0, \tag{8.127}$$

$$[A - \frac{1}{\tau}B_2 B_2^T Q(\tau)]R(\tau) + R(\tau)[A - \frac{1}{\tau}B_2 B_2^T Q(\tau)]^T \\ + B_1 B_1^T = 0. \tag{8.128}$$

Differentiating both sides of these equations with respect to τ, we obtain

$$[A - \frac{1}{\tau}B_2 B_2^T Q(\tau)]^T Q'(\tau) + Q'(\tau)[A - \frac{1}{\tau}B_2 B_2^T Q(\tau)] \\ + \frac{1}{\tau^2}Q(\tau)B_2 B_2^T Q(\tau) = 0, \tag{8.129}$$

$$[A - \frac{1}{\tau}B_2 B_2^T Q(\tau)]R'(\tau) + R'(\tau)[A - \frac{1}{\tau}B_2 B_2^T Q(\tau)]^T \\ + \frac{1}{\tau}B_2 B_2^T[\frac{1}{\tau}Q(\tau) - Q'(\tau)]R(\tau) + \frac{1}{\tau}R(\tau)[\frac{1}{\tau}Q(\tau) - Q'(\tau)]B_2 B_2^T = 0. \tag{8.130}$$

Subtracting (8.128) premultiplied by $Q(\tau)$ from (8.127) postmultiplied by $R(\tau)$, and taking the trace of the resulting equation, we obtain

$$\frac{1}{\tau}B_2^T Q(\tau)R(\tau)Q(\tau)B_2 + \text{tr}\{C_1 R(\tau)C_1^T\} - \text{tr}\{B_1^T Q(\tau)B_1\} = 0. \tag{8.131}$$

Differentiating both sides of this equation with respect to τ, we obtain

$$-\frac{1}{\tau^2}B_2^T Q(\tau)R(\tau)Q(\tau)B_2 - \text{tr}\{B_1^T Q'(\tau)B_1\} + \text{tr}\{C_1 R'(\tau)C_1^T\} \\ + \frac{1}{\tau}B_2^T[Q'(\tau)R(\tau)Q(\tau) + Q(\tau)R'(\tau)Q(\tau) + Q(\tau)R(\tau)Q'(\tau)]B_2 = 0. \tag{8.132}$$

Similarly, subtracting (8.130) premultiplied by $Q(\tau)$ from (8.127) postmultiplied by $R'(\tau)$, and taking the trace of the resulting equation, we obtain

$$\text{tr}\{C_1 R'(\tau)C_1^T\} - \frac{2}{\tau^2}B_2^T Q(\tau)R(\tau)Q(\tau)B_2 \\ + \frac{1}{\tau}B_2^T[Q'(\tau)R(\tau)Q(\tau) + Q(\tau)R'(\tau)Q(\tau) + Q(\tau)R(\tau)Q'(\tau)]B_2 = 0. \tag{8.133}$$

Thus, it follows from (8.132) and (8.133) that

$$f(\tau) = \text{tr}\{B_1^T Q'(\tau)B_1\}. \tag{8.134}$$

Differentiating both sides of this equation with respect to τ, we obtain

$$f'(\tau) = \text{tr}\{B_1^T Q''(\tau)B_1\}. \tag{8.135}$$

Moreover, differentiating both sides of (8.129) with respect to τ, we see that $Q''(\tau)$ satisfies the Lyapunov equation

$$[A - \frac{1}{\tau}B_2 B_2^T Q(\tau)]^T Q''(\tau) + Q''(\tau)[A - \frac{1}{\tau}B_2 B_2^T Q(\tau)]$$
$$- \frac{2}{\tau}[\frac{1}{\tau}Q(\tau) - Q'(\tau)]B_2 B_2^T [\frac{1}{\tau}Q(\tau) - Q'(\tau)] = 0. \tag{8.136}$$

Subtracting (8.129) from $1/\tau$ times (8.127), it also follows that

$$[A - \frac{1}{\tau}B_2 B_2^T Q(\tau)]^T [\frac{1}{\tau}Q(\tau) - Q'(\tau)] + [\frac{1}{\tau}Q(\tau) - Q'(\tau)][A - \frac{1}{\tau}B_2 B_2^T Q(\tau)]$$
$$+ \frac{1}{\tau}C_1^T C_1 = 0. \tag{8.137}$$

Equation (8.136) implies that $Q''(\tau)$ is negative semidefinite, and thus, $f'(\tau) \le 0$ for all $\tau \in (\rho, \infty)$. Therefore, $f'(N) \ge 0$, or equivalently, $f(N)$ is an increasing function of N. In addition, equation (8.137) implies that $Q'(\tau) - Q(\tau)/\tau$ is also negative semidefinite. Hence, since A has no eigenvalues in the open right-half plane, it follows that

$$\lim_{N \to 0^+} f(N) = \lim_{\tau \to \infty} f(\tau) = 0. \tag{8.138}$$

As a result, equation (8.123) has a unique solution for N. Note further that since R is positive semidefinite, the gain K is a global minimizer. $\qquad \square$

Proof of Corollary 6.1. This corollary follows from the fact that the right-hand side of (8.123) is strictly decreasing and the left-hand side is increasing, resulting in a unique intersection. $\qquad \square$

Proof of Theorem 6.2. We first show that the derivative of $\gamma^2(\rho)$ is nonnegative. For this purpose, define $\tau(\rho)$ as

$$\tau(\rho) = \frac{\rho + \lambda(\rho)}{N^2(\rho)}. \tag{8.139}$$

Then, it is clear that $\gamma^2(\rho)$ depends on ρ through $\tau(\rho)$. Thus, it follows from the chain rule that

$$\frac{d}{d\rho}\gamma^2(\rho) = \frac{d}{d\rho}\text{tr}\{C_1 R[\tau(\rho)]C_1^T\} = \frac{d}{d\tau}\text{tr}\{C_1 R(\tau)C_1^T\}\frac{d}{d\rho}\tau(\rho). \tag{8.140}$$

Proceeding similarly to the proof of Theorem 6.1, it can be shown that

$$\text{tr}\{C_1 R'(\tau)C_1^T\} = -\tau \text{tr}\{B_1^T Q''(\tau)B_1\}, \tag{8.141}$$

and $Q''(\tau)$ is negative semidefinite, and hence, $\text{tr}\{C_1 R'(\tau)C_1^T\} \ge 0$. The fact that $\tau'(\rho) > 0$ will be verified in Lemma 8.5 below, where it will be shown that $1/\tau(\rho)$ is a decreasing function of ρ. As a result, $\gamma^2(\rho)$ is an increasing function of ρ.

Next, we verify (6.15). Since $\gamma^2(\rho)$ is continuous and bounded from below, its limit as ρ goes to 0^+ exists. Moreover, it follows from (6.12) that $N^2\sigma_{\hat{u}}^2(\rho) \leq 2/\pi$ for all $\rho > 0$, and therefore,

$$\lim_{\rho \to 0^+} \gamma^2(\rho) \neq 0 \tag{8.142}$$

unless, of course, $C_1(sI - A)^{-1}B_1 = 0$. ☐

Proof of Corollary 6.2. This corollary follows from the fact that $\text{Prob}\{|\hat{u}| > 1\} = 1 - \text{Prob}\{|\hat{u}| \leq 1\}$ and $\text{Prob}\{|\hat{u}| \leq 1\} = N$. ☐

Proof of Theorem 6.3. *Part (i):* By definition, any point $x_G \in \mathbf{R}^{n_x}$ is the equilibrium point of (6.17) if

$$Ax_G + B_2\varphi(Kx_G) = 0, \tag{8.143}$$

or, equivalently,

$$(A + B_2K)x_G + B_2[\varphi(Kx_G) - Kx_G] = 0. \tag{8.144}$$

Since $A + B_2K$ is nonsingular, it follows that

$$x_G + (A + B_2K)^{-1}B_2[\varphi(Kx_G) - Kx_G] = 0. \tag{8.145}$$

Premultiplying both sides of this equation by K, we obtain

$$[1 - K(A + B_2K)^{-1}B_2]Kx_G + K(A + B_2K)^{-1}B_2\varphi(Kx_G) = 0. \tag{8.146}$$

If $K(A + B_2K)^{-1}B_2 = 0$, then this equation implies that $Kx_G = 0$, and in turn, equation (8.145) implies that $x_G = 0$. If, on the other hand, $K(A + B_2K)^{-1}B_2 = 1$, then equation (8.146) implies that $Kx_G = 0$, and then, equation (8.145) implies that $x_G = 0$. Hence, assume that $K(A + B_2K)^{-1}B_2 \neq 0$ and $K(A + B_2K)^{-1}B_2 \neq 1$, then we claim that

$$K(A + B_2K)^{-1}B_2 < 1. \tag{8.147}$$

To see this, note that

$$\begin{aligned}1 - K(A + B_2K)^{-1}B_2 &= \det[I - (A + B_2K)^{-1}B_2K] \\ &= \det[(A + B_2K)^{-1}]\det(A) \geq 0,\end{aligned} \tag{8.148}$$

and by assumption $K(A + B_2K)^{-1}B_2 \neq 1$. Thus, under the above assumptions, it follows from (8.147) that either

$$\frac{K(A + B_2K)^{-1}B_2 - 1}{K(A + B_2K)^{-1}B_2} < 0, \tag{8.149}$$

or

$$\frac{K(A + B_2K)^{-1}B_2 - 1}{K(A + B_2K)^{-1}B_2} > 1. \tag{8.150}$$

Hence, rewriting (8.146) as

$$\varphi(Kx_G) = \frac{K(A+B_2K)^{-1}B_2 - 1}{K(A+B_2K)^{-1}B_2} Kx_G, \tag{8.151}$$

we see that x_G satisfies this equation only when $Kx_G = 0$. Hence, once again equation (8.145) implies that $x_G = 0$.

Part (ii): The fact that the origin $x_G = 0$ is the unique exponentially stable equilibrium of the closed loop system follows from the Lyapunov's indirect method.

Part (iii): Since A has no eigenvalues with positive real part, (A, B_2) is stabilizable and (C_1, A) is detectable, letting $\{A_{co}, B_{2co}, C_{1co}, D_{12co}\}$ be the controllable and observable part of the Kalman canonical decomposition of $\{A, B_2, C_1, D_{12}\}$, it is straightforward to see that the system (6.17) is asymptotically stable if and only if the system

$$\begin{aligned}
\dot{x}_{G_{co}} &= A_{co}x_{G_{co}} + B_{2co}\varphi(u), \\
z_{co} &= C_{1co}x_{G_{co}} + D_{12co}u_{co}, \\
u_{co} &= K_{co}x_{G_{co}}
\end{aligned} \tag{8.152}$$

is asymptotically stable. Thus, without loss of generality, we assume that (A, B_2) is controllable and (C_1, A) is observable. Then, it follows from the Riccati equation (6.10) that Q is positive definite. Hence, with ε as defined in (6.19), consider the Lyapunov function candidate

$$V(x_G) = x_G^T(\varepsilon Q)x_G. \tag{8.153}$$

Taking the derivative of $V(x_G)$ along the trajectories of (6.17), we obtain

$$\begin{aligned}
\dot{V}(x_G) &= -x_G^T[A^T(\varepsilon Q) + (\varepsilon Q)A]x_G + 2x_G^T(\varepsilon Q)B_2\varphi(u) \\
&= -x_G^T(\varepsilon C_1^T C_1)x_G - \varepsilon(\rho + \lambda)[2/Nu\varphi(u) - u^2].
\end{aligned} \tag{8.154}$$

Thus, $\dot{V}(x_G) \le 0$ if

$$|u| \le \frac{2}{N}, \tag{8.155}$$

or equivalently,

$$|B_2^T(\varepsilon Q)x_G| \le 2. \tag{8.156}$$

Now, define the set $\mathcal{X}$ as

$$\mathcal{X} = \{x_G \in \mathbf{R}^{n_{x_G}} : x_G^T(\varepsilon Q)x_G \le \gamma^2\}, \tag{8.157}$$

and select γ sufficiently small so that the set $\mathcal{U}$ defined as

$$\mathcal{U} = \{x_G \in \mathbf{R}^{n_{x_G}} : x_G^T(\varepsilon Q)B_2B_2^T(\varepsilon Q)x_G \le 4\} \tag{8.158}$$

contains $\mathcal{X}$. Moreover, define $\mathcal{S}$ as

$$\mathcal{S} = \{x_G \in \mathcal{X} : \dot{V}(x_G) = 0\} \tag{8.159}$$

and suppose that x_G is a trajectory that belongs identically to S. Then, it follows from (8.154) that $C_1 x_G = 0$ and $K x_G = 0$. Since (C_1, A) is observable, x_G must be the trivial trajectory $x_G = 0$. Hence, by the LaSalle theorem, the equilibrium point $x_G = 0$ of the closed loop system (6.17) is asymptotically stable. Moreover, the set X defined above is a subset of the domain of attraction of the closed loop system. To verify that X defined in (6.18) is a subset of the domain of attraction of the closed loop system, we will find the best possible γ such that $U \supset X$. The coordinate transformation $\bar{x}_G = \varepsilon^{1/2} Q^{1/2} x_G$ transforms X and U into

$$\bar{X} = \{\bar{x}_G \in \mathbf{R}^{n_{x_G}} : \bar{x}_G^T \bar{x}_G \le \gamma^2\}, \tag{8.160}$$

and

$$\bar{U} = \{\bar{x}_G \in \mathbf{R}^{n_{x_G}} : \bar{x}_G^T (\varepsilon^{1/2} Q^{1/2}) B_2 B_2^T (\varepsilon^{1/2} Q^{1/2}) \bar{x}_G \le 4\}, \tag{8.161}$$

respectively. Thus, $U \supset X$ if and only if $\bar{U} \supset \bar{X}$, or equivalently, the radius r of the hypersphere $\bar{x}_G^T \bar{x}_G = \gamma^2$ is less than or equal to the distance d between the origin $\bar{x}_G = 0$ and the hyperplane $B_2^T \varepsilon^{1/2} Q^{1/2} \bar{x}_G = 2$ (see Figure 8.2). Clearly, $r = \gamma$ and it can be easily shown that d is given by

$$d^2 = \frac{4}{B_2^T (\varepsilon Q) B_2}. \tag{8.162}$$

Thus, $r \le d$ if

$$\gamma^2 \le \frac{4}{B_2^T (\varepsilon Q) B_2}. \tag{8.163}$$

Hence, the set X is a subset of the domain of attraction of the closed loop system.

Part (iv): The solution of (6.17) starting from $x_G(0)$ satisfies

$$x_G(t) = \exp(At)x_G(0) + \int_0^t \exp[A(t - \tau)] B_2 \varphi[K x_G(\tau)] d\tau. \tag{8.164}$$

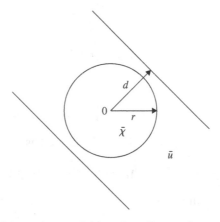

Figure 8.2. Illustration of the estimate of the region of attraction.

Hence,

$$\|x_G(t)\| \leq \|\exp(At)x_G(0)\| + \|\int_0^t \exp[A(t-\tau)]B_2\varphi[Kx_G(\tau)]d\tau\|. \qquad (8.165)$$

Then, it follows that

$$\|x_G(t)\| \leq \exp(-\nu t)\|x_G(0)\| + \frac{1}{\nu}[1 - \exp(-\nu t)]\|B_2\| $$
$$\leq \|x_G(0)\| + \frac{1}{\nu}\|B_2\|, \qquad (8.166)$$

where ν is the real part of the rightmost eigenvalue of A. Thus, letting

$$m = \|x_G(0)\| + \frac{1}{\nu}\|B_2\|, \qquad (8.167)$$

we see that $\|x_G(t)\| \leq m$ for all $t \geq 0$.   □

To prove Theorem 6.4, we need the following two lemmas:

Lemma 8.4. *Assume that (A, B_2) is stabilizable, (C_1, A) is detectable and A has no eigenvalues in the open right-half plane. Then, for any $\varepsilon > 0$, there exists a unique positive semidefinite matrix $Q(\varepsilon)$ that satisfies*

$$A^T Q(\varepsilon) + Q(\varepsilon)A - \varepsilon Q(\varepsilon)B_2 B_2^T Q(\varepsilon) + C_1^T C_1 = 0 \qquad (8.168)$$

such that $A - B_2 B_2^T \varepsilon Q(\varepsilon)$ is Hurwitz. Moreover,

$$\lim_{\varepsilon \to 0^+} \varepsilon Q(\varepsilon) = 0. \qquad (8.169)$$

In addition, if (C_1, A) is not only detectable but observable, then $Q(\varepsilon)$ is positive definite for all $\varepsilon > 0$.

Proof of Lemma 8.4. The existence of a unique positive semidefinite solution $Q(\varepsilon)$ for all $\varepsilon > 0$ is a standard result. Moreover, it is also easy to establish that $Q(\varepsilon)$ is positive definite if (C_1, A) has no stable unobservable modes. Multiplying both sides of equation (8.168) by ε and letting $\bar{Q}(\varepsilon) = \varepsilon Q(\varepsilon)$, we obtain

$$A^T \bar{Q}(\varepsilon) + \bar{Q}(\varepsilon)A - \bar{Q}(\varepsilon)B_2 B_2^T \bar{Q}(\varepsilon) + \varepsilon C_1^T C_1 = 0. \qquad (8.170)$$

When $\varepsilon = 0$, the only solution to this equation, for which all eigenvalues of $A - B_2 B_2^T \bar{Q}(\varepsilon)$ are in the closed left-half plane, is $\bar{Q}(\varepsilon) = 0$. On the other hand, it is relatively easy to see that the solution of this Riccati equation is continuous at $\varepsilon = 0$ from the right. Hence, $\bar{Q}(\varepsilon)$ approaches 0 as $\varepsilon \to 0^+$. Since $\bar{Q}(\varepsilon) = \varepsilon Q(\varepsilon)$, (8.169) follows.   □

Lemma 8.5. *With $\rho > 0$ being a parameter, assume that A, B_1, B_2, C_1, and D_{12} satisfy Assumption 6.1. For each ρ, let $(N(\rho), Q(\rho), R(\rho), \lambda(\rho))$ be the unique solution of the system of equations (6.10)–(6.13) and define $\varepsilon(\rho)$ as*

$$\varepsilon(\rho) = \frac{N^2(\rho)}{\rho + \lambda(\rho)}. \qquad (8.171)$$

Then, $\varepsilon(\rho)$ is a (strictly) decreasing function of ρ and

$$\lim_{\rho \to \infty} \varepsilon(\rho) = 0^+. \tag{8.172}$$

Moreover, $\varepsilon(\rho)$ is also continuous.

Proof of Lemma 8.5. With $\varepsilon(\rho)$ as defined in (8.171), omitting the explicit dependence of $\varepsilon(\rho)$ on ρ, equations (6.10)–(6.13) can be rewritten as

$$\varepsilon - \frac{\sqrt{\pi} N^2 \exp\left(\left[\mathrm{erf}^{-1}(N)\right]^2\right) - 2N \mathrm{erf}^{-1}(N)}{\sqrt{\pi} \exp\left(\left[\mathrm{erf}^{-1}(N)\right]^2\right)} \frac{1}{\rho} = 0, \tag{8.173}$$

$$A^T Q + QA - \varepsilon Q B_2 B_2^T Q + C_1^T C_1 = 0, \tag{8.174}$$

$$(A - \varepsilon B_2 B_2^T Q)R + R(A - \varepsilon B_2 B_2^T Q)^T + B_1 B_1^T = 0, \tag{8.175}$$

$$\varepsilon^2 B_2^T Q R Q B_2 - \frac{N^2 \alpha^2}{2\left[\mathrm{erf}^{-1}(N)\right]^2} = 0. \tag{8.176}$$

Clearly, $\varepsilon(\rho)$ is a continuous function of ρ. Noting that Q and R are functions of ε, define

$$f(\varepsilon) = \varepsilon^2 B_2^T Q R Q B_2,$$

$$g(N) = \frac{\sqrt{\pi} N^2 \exp\left(\left[\mathrm{erf}^{-1}(N)\right]^2\right) - 2N \mathrm{erf}^{-1}(N)}{\sqrt{\pi} \exp\left(\left[\mathrm{erf}^{-1}(N)\right]^2\right)}, \tag{8.177}$$

$$h(N) = \frac{N^2 \alpha^2}{2\left[\mathrm{erf}^{-1}(N)\right]^2}.$$

Then, equations (8.173) and (8.176) can be written as

$$\begin{aligned} f_1(\varepsilon, N, \rho) &= \varepsilon - g(N)/\rho = 0, \\ f_1(\varepsilon, N, \rho) &= f(\varepsilon) - h(N) = 0. \end{aligned} \tag{8.178}$$

Hence, by the implicit function theorem, it follows that

$$\frac{d\varepsilon}{d\rho} = -\frac{\dfrac{\delta(f_1, f_2)}{\delta(\rho, N)}}{\dfrac{\delta(f_1, f_2)}{\delta(\varepsilon, N)}} = \frac{\dfrac{1}{\rho^2} g(N) h'(N)}{\dfrac{1}{\rho} f'(\varepsilon) g'(N) - h'(N)}. \tag{8.179}$$

It is straightforward to verify that, for all $N \in (0,1)$, $g(N) > 0$, $g'(N) < 0$, and $h'(N) < 0$. Moreover, it follows from the proof of Theorem 6.1 that $f'(\varepsilon) \geq 0$ for all $\varepsilon > 0$. Hence, $\varepsilon'(\rho) < 0$, and this implies that $\varepsilon(\rho)$ is a decreasing function of ρ. In addition, proceeding similarly, it can be shown that

$$\frac{dN}{d\rho} = \frac{\frac{1}{\rho^2}f'(\varepsilon)g(N)}{\frac{1}{\rho}f'(\varepsilon)g'(N) - h'(N)}, \tag{8.180}$$

which implies that $N(\rho)$ is an increasing function of ρ. Next, we show that as $\rho \to \infty$, $\varepsilon(\rho)$ approaches 0^+. Since $\rho(\varepsilon)$ is continuous, strictly decreasing and bounded from below by 0, the limit in (8.172) exists. Since $N(\rho)$ is an increasing function of ρ and is not identically equal to zero, it follows from (8.173) that as $\rho \to \infty$, $\varepsilon(\rho)$ approaches 0^+. Moreover, it follows from (8.176) that as $\rho \to \infty$, $N(\rho)$ approaches 1^-.   □

Proof of Theorem 6.4. It follows from Theorem 6.3 and Lemma 8.4 that, for each $\rho > 0$, the SLQR state feedback control law parameterized by ρ guarantees that the equilibrium $x_G = 0$ of the undisturbed closed loop system (6.17) is asymptotically stable and the set

$$\mathcal{X}(\rho) = \left\{ x_G \in \mathbf{R}^{n_{x_G}} : x_G^T(\varepsilon(\rho)Q[\varepsilon(\rho)])x_G \leq \frac{4}{B_2^T(\varepsilon(\rho)Q[\varepsilon(\rho)])B_2} \right\} \tag{8.181}$$

is a subset of its domain of attraction. Since, by Lemma 8.5, $\varepsilon(\rho)$ is a decreasing function of ρ, it follows that the family of ellipsoidal sets $\mathcal{X}(\rho)$ is an increasing nested set family, that is, $\mathcal{X}(\rho_1) \subset \mathcal{X}(\rho_2)$ whenever $\rho_1 < \rho_2$. Moreover, since $\varepsilon(\rho)$ approaches 0^+ as $\rho \to \infty$, it follows that

$$\lim_{\rho \to \infty} \mathcal{X}(\rho) = \mathbf{R}^{n_{x_G}}. \tag{8.182}$$

Hence, for the given bounded set $\mathcal{B}$, letting ρ^* be the smallest ρ such that $\mathcal{X}(\rho) \supset \mathcal{B}$, we conclude that, for every $\rho \geq \rho^*$, the SLQR state feedback control law parameterized by ρ guarantees that the equilibrium $x_G = 0$ of the undisturbed closed loop system is asymptotically stable and $\mathcal{B}$ is contained in its region of attraction.   □

Proof of Theorem 6.5. Let Ω be the set $\Omega = \{(K,L,M) : \tilde{A}$ is Hurwitz$\}$, where $\tilde{A}$ is as defined in (6.29). Then, for any $(K,L,M) \in \Omega$,

$$\sigma_{\tilde{z}}^2 = \text{tr}\{\tilde{C}\tilde{P}\tilde{C}^T\}, \tag{8.183}$$

where $(\tilde{P},N)$ satisfies

$$\tilde{A}\tilde{P} + \tilde{P}\tilde{A}^T + \tilde{B}\tilde{B}^T = 0, \tag{8.184}$$

$$\tilde{K}\tilde{P}\tilde{K}^T - \frac{1}{2\left[\text{erf}^{-1}(N)\right]^2} = 0. \tag{8.185}$$

Hence, to within a similarity transformation, the problem at hand is equivalent to

$$\underset{(K,L,M)\in\Omega}{\text{minimize}}\,\text{tr}\{\tilde{C}\tilde{P}\tilde{C}^T\},$$

$$\text{subject to}\,\tilde{A}\tilde{P}+\tilde{P}\tilde{A}^T+\tilde{B}\tilde{B}^T=0, \tag{8.186}$$

$$\tilde{K}\tilde{P}\tilde{K}^T-\frac{1}{2\big[\text{erf}^{-1}(N)\big]^2}=0.$$

Again, we use the Lagrange multiplier method to solve this constrained optimization problem and proceed similarly to the proof of Theorem 6.1.

First, we verify the regularity of the constraints. Let $(K,L,M,N,\tilde{P})$ be such that it satisfies the constraints (8.184) and (8.185), and for an arbitrary symmetric matrix $\tilde{Q}$ and real number λ, define the function

$$\Phi(M,L,K,N,\tilde{P})=\text{tr}\{[\tilde{A}\tilde{P}+\tilde{P}\tilde{A}^T+\tilde{B}\tilde{B}^T]\tilde{Q}\}+\lambda\left(\tilde{K}\tilde{P}\tilde{K}^T-\frac{1}{2\big[\text{erf}^{-1}(N)\big]^2}\right). \tag{8.187}$$

Then with $\tilde{P}$ and $\tilde{Q}$ partitioned as

$$\tilde{P}=\left[\begin{array}{cc}P_{11} & P_{12}^T\\ P_{12} & P_{22}\end{array}\right],\quad \tilde{Q}=\left[\begin{array}{cc}Q_{11} & Q_{12}\\ Q_{12}^T & Q_{22}\end{array}\right], \tag{8.188}$$

we have

$$\frac{\partial\Phi}{\partial K}=0\Rightarrow NB_2^T(Q_{11}P_{12}^T+Q_{12}P_{22})+\lambda KP_{22}=0, \tag{8.189}$$

$$\frac{\partial\Phi}{\partial L}=0\Rightarrow Q_{22}L\mu-(Q_{12}^TP_{11}+Q_{22}P_{12})C_2^T=0, \tag{8.190}$$

$$\frac{\partial\Phi}{\partial M}=0\Rightarrow Q_{12}^TP_{12}^T+Q_{22}P_{22}=0, \tag{8.191}$$

$$\frac{\partial\Phi}{\partial N}=0\Rightarrow K(P_{12}Q_{11}+P_{22}Q_{12}^T)B_2+\lambda\frac{\sqrt{\pi}}{4}\frac{\exp\left(\big[\text{erf}^{-1}(N)\big]^2\right)}{\big[\text{erf}^{-1}(N)\big]^3}=0, \tag{8.192}$$

$$\frac{\partial\Phi}{\partial\tilde{P}}=0\Rightarrow \tilde{A}^T\tilde{Q}+\tilde{Q}\tilde{A}+\lambda\tilde{K}^T\tilde{K}=0. \tag{8.193}$$

Multiplying (8.189) from right by K^T and using (8.192), (8.185) we obtain

$$\lambda\left[\frac{1}{2\big[\text{erf}^{-1}(N)\big]^2}-\frac{\sqrt{\pi}}{4}\frac{N}{\big[\text{erf}^{-1}(N)\big]^3}\exp\left(\big[\text{erf}^{-1}(N)\big]^2\right)\right]=0. \tag{8.194}$$

Here, since the expression that multiplies λ is nonzero for all $N\in(0,1)$ and as $N\to 0^+$, $N\to 1^-$, we conclude that $\lambda=0$. Then, with $\lambda=0$, it follows from (8.193) that $\tilde{Q}=0$. Hence, the regularity conditions are satisfied.

Next, we form the Lagrangian

$$\Psi(K,L,M,N,\tilde{P},\tilde{Q},\lambda) = \mathrm{tr}\{\tilde{C}\tilde{P}\tilde{C}^T\} + \mathrm{tr}\{[\tilde{A}\tilde{P} + \tilde{P}\tilde{A}^T + \tilde{B}\tilde{B}^T]\tilde{Q}\}$$

$$+ \lambda\left(\tilde{K}\tilde{P}\tilde{K}^T - \frac{1}{2\left[\mathrm{erf}^{-1}(N)\right]^2}\right), \tag{8.195}$$

where $\tilde{Q}$ and λ are the Lagrange multipliers. Differentiating Ψ with respect to K, L, M, N, $\tilde{P}$, $\tilde{Q}$, and λ, and equating the results to zero, we obtain

$$(\rho + \lambda)KP_{22} + NB_2^T(Q_{11}P_{12}^T + Q_{12}P_{22}) = 0, \tag{8.196}$$

$$Q_{22}L\mu - (Q_{12}^T P_{11} + Q_{22}P_{12})C_2^T = 0, \tag{8.197}$$

$$Q_{12}^T P_{12}^T + Q_{22}P_{22} = 0, \tag{8.198}$$

$$K(P_{12}Q_{11} + P_{22}Q_{12}^T)B_2 + \lambda\frac{\sqrt{\pi}}{4}\frac{\exp\left(\left[\mathrm{erf}^{-1}(N)\right]^2\right)}{\left[\mathrm{erf}^{-1}(N)\right]^3} = 0, \tag{8.199}$$

$$\tilde{A}^T\tilde{Q} + \tilde{Q}\tilde{A} + \tilde{C}^T\tilde{C} + \lambda\tilde{K}^T\tilde{K} = 0, \tag{8.200}$$

$$\tilde{A}\tilde{P} + \tilde{P}\tilde{A}^T + \tilde{B}\tilde{B}^T = 0, \tag{8.201}$$

$$KP_{22}K^T - \frac{1}{2\left[\mathrm{erf}^{-1}(N)\right]^2} = 0, \tag{8.202}$$

where we have used the partitions

$$\tilde{P} = \begin{bmatrix} P_{11} & P_{12}^T \\ P_{12} & P_{22} \end{bmatrix}, \quad \tilde{Q} = \begin{bmatrix} Q_{11} & Q_{12} \\ Q_{12}^T & Q_{22} \end{bmatrix}. \tag{8.203}$$

Now, assuming that P_{22} and Q_{22} are nonsingular, equations (8.196) and (8.197) yield

$$K = -\frac{N}{\rho + \lambda}B_2^T(Q_{11}P_{12}^T + Q_{12}P_{22})P_{22}^{-1}, \tag{8.204}$$

and

$$L = Q_{22}^{-1}(Q_{12}^T P_{11} + Q_{22}P_{12})C_2^T\frac{1}{\mu}, \tag{8.205}$$

respectively. Moreover, it follows from (8.198) that

$$-Q_{22}^{-1}Q_{12}^T P_{12}^T P_{22}^{-1} = I. \tag{8.206}$$

Then, letting $T = P_{12}^T P_{22}^{-1}$, it follows from equation (8.204) that

$$K = -\frac{N}{\rho + \lambda}B_2^T(Q_{11}P_{12}^T P_{22}^{-1} + Q_{12})$$

$$= -\frac{N}{\rho + \lambda}B_2^T(Q_{11}P_{12}^T P_{22}^{-1} - Q_{12}Q_{22}^{-1}Q_{12}^T P_{12}^T P_{22}^{-1}) \tag{8.207}$$

$$= -\frac{N}{\rho + \lambda}B_2^T(Q_{11} - Q_{12}Q_{22}^{-1}Q_{12}^T)T.$$

Similarly, noting $T^{-1} = -Q_{22}^{-1}Q_{12}^T$, it follows from equation (8.205) that

$$
\begin{aligned}
L &= (Q_{22}^{-1}Q_{12}^T P_{11} + P_{12})C_2^T \frac{1}{\mu} \\
&= (Q_{22}^{-1}Q_{12}^T P_{11} - Q_{22}^{-1}Q_{12}^T P_{12}^T P_{22}^{-1} P_{12})C_2^T \frac{1}{\mu} \\
&= -T^{-1}(P_{11} - P_{12}^T P_{22}^{-1} P_{12})C_2^T \frac{1}{\mu}.
\end{aligned}
\tag{8.208}
$$

Hence, defining P and Q as

$$
P = P_{11} - P_{12}^T P_{22}^{-1} P_{12}, \quad Q = Q_{11} - Q_{12}Q_{22}^{-1}Q_{12}^T,
\tag{8.209}
$$

we obtain

$$
K = -\frac{N}{\rho + \lambda}B_2^T QT,
\tag{8.210}
$$

and

$$
L = -T^{-1}PC_2^T \frac{1}{\mu}.
\tag{8.211}
$$

In addition, equations (8.200) and (8.201) can be rewritten as

$$
A^T Q_{11} + Q_{11}A - C_2^T L^T Q_{12}^T - Q_{12}LC_2 + C_1^T C_1 = 0,
\tag{8.212}
$$

$$
A^T Q_{12} + Q_{12}M - C_2^T L^T Q_{22} + Q_{11}B_2 NK = 0,
\tag{8.213}
$$

$$
M^T Q_{12}^T + Q_{12}^T A - Q_{22}LC_2 + K^T NB_2^T Q_{11} = 0,
\tag{8.214}
$$

$$
M^T Q_{22} + Q_{22}M + K^T NB_2^T Q_{12} + Q_{12}^T B_2 NK + (\rho + \lambda)K^T K = 0,
\tag{8.215}
$$

$$
AP_{11} + P_{11}A^T + B_2 NKP_{12} + P_{12}^T K^T NB_2^T + B_1 B_1^T = 0,
\tag{8.216}
$$

$$
MP_{12} + P_{12}A^T + P_{22}K^T NB_2^T - LC_2 P_{11} = 0,
\tag{8.217}
$$

$$
AP_{12}^T + P_{12}^T M^T + B_2 NKP_{22} - P_{11}C_2^T L^T = 0,
\tag{8.218}
$$

$$
MP_{22} + P_{22}M^T - LC_2 P_{12}^T - P_{12}C_2^T L^T + LL^T \mu = 0.
\tag{8.219}
$$

Premultiplying both sides of (8.218) by T^{-1} and subtracting the resulting equation from (8.219) yields

$$
\begin{aligned}
MP_{22} - T^{-1}AP_{12}^T - T^{-1}B_2 NKP_{22} - LC_2 P_{12}^T \\
+P_{22}M^T - T^{-1}P_{12}^T M^T + T^{-1}P_{11}C_2^T L^T + LL^T \mu - P_{12}C_2^T L^T = 0.
\end{aligned}
\tag{8.220}
$$

Simplifying the left-hand side of this equation we obtain

$$
[M - T^{-1}(A + B_2 NKT^{-1} + TLC_2)T]P_{22} = 0.
\tag{8.221}
$$

Thus, it follows that

$$
M = T^{-1}(A + B_2 NKT^{-1} + TLC_2)T.
\tag{8.222}
$$

Having obtained K, L, and M, we verify equations (6.32), (6.33), and (6.34)–(6.39). To this end, define

$$R = P_{12}^T P_{22}^{-1} P_{12}, \quad S = Q_{12} Q_{22}^{-1} Q_{12}^T. \tag{8.223}$$

Premultiplying (8.217) by T and substituting K, L, and M into the resulting equation gives the Lyapunov equation (6.36). Similarly, postmultiplying (8.213) by T^{-1} and substituting K, L, and M into the resulting equation gives the Lyapunov equation (6.37). Substituting K, L, and M into (8.216) and (8.212), and using (6.36) and (6.37) gives the Riccati equations (6.34) and (6.35), respectively. Multiplying (8.196) from right by K^T and using (8.199), (8.202) gives (6.39). Substituting K into (8.202) and using the definition of R gives (6.38). Moreover, assuming that the triple (K, L, M), given above, is in Ω and substituting it into (8.183) gives (6.32). To verify (6.33), we apply the similarity transformation T^{-1} to the state space realization $\{M, L, K\}$ to get $\{\bar{M}, \bar{L}, \bar{K}\} = \{TMT^{-1}, TL, KT^{-1}\}$. Hence, it follows that

$$\bar{K} = -\frac{N}{\rho + \lambda} B_2^T Q,$$

$$\bar{L} = -PC_2^T \frac{1}{\mu}, \tag{8.224}$$

$$\bar{M} = A + B_2 N \bar{K} + \bar{L} C_2,$$

which are identical to the equations in (6.33), except that for notational simplicity we have dropped the bars in writing the equations in (6.33).

Finally, we show that the system of equations (6.34)–(6.39) has a unique solution for (N, P, Q, R, S, λ) such that $(\bar{K}, \bar{L}, \bar{M}) \in \Omega$. Since (A, B_1) is stabilizable and (C_2, A) is detectable, the Riccati equation (6.34) has a unique positive semidefinite solution for P such that $A + \bar{L} C_2$ is Hurwitz. Moreover, note that P does not depend on N, Q, R, S, and λ. With this P, since (A, B_2) is stabilizable and (C_1, A) is detectable, it follows from the proof of Theorem 6.1 that equations (6.39), (6.35), (6.36), and (6.38) have a unique solution (N, Q, R, λ) such that $A + B_2 N \bar{K}$ is Hurwitz, and Q and R are positive semidefinite. Then substituting P and Q into (6.37) and noting that $A + \bar{L} C_2$ is Hurwitz, it follows that the Lyapunov equation (6.37) has a unique positive semidefinite solution for S. In addition, since $A + B_2 N \bar{K}$ and $A + \bar{L} C_2$ are Hurwitz, so is $\tilde{A}$. Hence, equations (6.34)–(6.39) have a unique solution for (N, P, Q, R, S, λ) such that $(\bar{K}, \bar{L}, \bar{M}) \in \Omega$. Moreover, since $(\bar{K}, \bar{L}, \bar{M})$ that minimizes the performance measure is unique within a similarity transformation, we conclude that $C(s) = \bar{K}(sI - \bar{M})^{-1} \bar{L}$ is also unique.

In the above derivations, we have assumed that P_{22} and Q_{22} are nonsingular. However, by using pseudoinverses it can be shown that this assumption is not necessary. In fact, it can be shown that, P_{22} is nonsingular if and only if R is nonsingular, and, similarly, that Q_{22} is nonsingular if and only if S is nonsingular. When R is singular, any $\bar{K}_n$ in the left null space of R can be added to $\bar{K}$ in (8.224) without affecting the value of the performance measure, provided that this does not destroy internal stability. Similarly, when S is singular, any $\bar{L}_n$ in the right null space of S can be added to $\bar{L}$ in (8.224) without affecting the value of the performance measure, provided that this does not destroy internal stability. Thus, if R or S is singular, then, apart from a similarity transformation, $(\bar{K},\bar{L},\bar{M})$ that minimizes the performance measure is not unique. Moreover, noting that R is nonsingular if and only if $(A + B_2 N\bar{K}, PC_2^T)$ is controllable, and that S is nonsingular if and only if $(B_2^T Q, A + \bar{L}C_2)$ is observable, we see that this nonuniqueness occurs if and only if the realization $\{\bar{M},\bar{L},\bar{K}\}$ is not minimal. Hence, after cancelling the poles and zeros that correspond to such uncontrollable or unobservable modes, we conclude that the resulting controller $C(s)$ is still unique. Moreover, it follows from this discussion that the realization $\{\bar{M},\bar{L},\bar{K}\}$ is minimal if R and S are nonsingular. Note further that since $\tilde{P}$ is positive semidefinite, $(\bar{K},\bar{L},\bar{M})$ is a global minimizer. $\qquad\square$

Proof of Theorem 6.6. The proof of this theorem is analogous to the proof of Theorem 6.3, but using the Lyapunov function

$$V(x_G, x_C) = x_G^T Q x_G + (x_G - x_C)^T S(x_G - x_C). \tag{8.225}$$

$\qquad\square$

Proof of Theorem 6.7. We again use the Lagrange multiplier method to find the necessary conditions for optimality. First, the regularity of the constraints is verified. Let (K, N, R, α) satisfy (6.68) and (6.69), and for an arbitrary symmetric matrix Q and real number λ define

$$\Phi(K, N, R, \alpha) = \text{tr}\left(\left[(A + B_2 NK)R + R(A + B_2 NK)^T + B_1 B_1^T\right]Q\right)$$

$$+ \lambda\left(KRK^T - \frac{\alpha^2}{2\left(\text{erf}^{-1}(N)\right)^2}\right). \tag{8.226}$$

Differentiating Φ with respect to K, N, R, α and equating to zero, we obtain:

$$NB_2^T QR + \lambda KR = 0, \tag{8.227}$$

$$KRQB_2 + \lambda\frac{\sqrt{\pi}}{4}\alpha^2\frac{\exp\left(\text{erf}^{-1}(N)^2\right)}{\text{erf}^{-1}(N)^3} = 0, \tag{8.228}$$

$$(A + B_2 NK)^T Q + Q(A + B_2 NK) + \lambda K^T K = 0, \tag{8.229}$$

$$\frac{-2\lambda\alpha}{2\text{erf}^{-1}(N)^2} = 0. \tag{8.230}$$

Since $N \in (0,1)$ and $\alpha > 0$, it follows from (8.230) that $\lambda = 0$, which, through (8.229), implies that $Q = 0$. Consequently, (8.227) and (8.228) are satisfied and the constraints are regular.

Next, we form the Lagrangian:

$$\Psi(K,N,R,Q,\lambda,\alpha) = \mathrm{tr}\left(C_1 R C_1^T\right) + \rho K R K^T + \eta \alpha^2$$

$$+ \mathrm{tr}\left(\left[(A+B_2 NK)R + R(A+B_2 NK)^T + B_1 B_1^T\right]Q\right) + \lambda\left(KRK^T - \frac{\alpha^2}{2\left(\mathrm{erf}^{-1}(N)\right)^2}\right). \tag{8.231}$$

Differentiating Ψ with respect to K,N,R,Q,λ,α results in

$$\left((\rho+\lambda)K + NB_2^T Q\right)R = 0, \tag{8.232}$$

$$KRQB_2 + \lambda\alpha^2 \frac{\sqrt{\pi}}{4} \frac{\exp\left(\mathrm{erf}^{-1}(N)^2\right)}{\left(\mathrm{erf}^{-1}(N)\right)^3} = 0, \tag{8.233}$$

$$(A+B_2 NK)^T Q + Q(A+B_2 NK) + C_1^T C_1 + (\rho+\lambda)K^T K = 0, \tag{8.234}$$

$$(A+B_2 NK)R + R(A+B_2 NK)^T + B_1 B_1^T = 0, \tag{8.235}$$

$$KRK^T - \frac{\alpha^2}{2\mathrm{erf}^{-1}(N)^2} = 0, \tag{8.236}$$

$$2\eta\alpha - \frac{2\lambda\alpha}{2\mathrm{erf}^{-1}(N)^2} = 0. \tag{8.237}$$

The equations (6.70) and (6.71) for the parameters K and α follow immediately from (8.232) and (8.236), respectively. Substituting (6.70) into (8.234) and (8.235) yielding (6.72) and (6.73). Multiplying (8.232) from the right by K^T and using (8.233) and (8.236) yields (6.74). Finally, (6.75) follows immediately from (8.237).

We now argue that the ILQR problem (6.66) has a solution, that is, the minimum exists and is attained. For that purpose, note that (6.66) can be reformulated as

$$\min_{\alpha}\left\{\eta\alpha^2 + \min_{K}\sigma_{\tilde{z}}^2\right\}. \tag{8.238}$$

It is known from SLQR theory (Section 6.1) that, under Assumption 6.1, for every $\alpha > 0$, the gain K that solves the minimization problem

$$\min_{K}\sigma_{\tilde{z}}^2 \tag{8.239}$$

exists. Moreover, the achieved minimum is continuous with respect to α. Thus, the function

$$y(\alpha) = \eta\alpha^2 + \min_K \upsilon_z^2 \qquad (8.240)$$

is continuous for $\alpha \geq 0$ and tends to ∞ as $\alpha \to \infty$. Hence, $q(\alpha)$ achieves a minimum at some $a^* \geq 0$. Let K^* be the minimizer of (8.239) when $\alpha = \alpha^*$. Then the pair (K^*, α^*) solves (6.66).

Finally, we prove that (6.72)–(6.75) admits a unique solution. Since (A, B_2) is stabilizable and (C_1, A) is detectable, for any $N \in (0,1), \lambda > 0$, the Ricatti equation (6.72) has a unique positive semidefinite solution Q such that $(A + B_2 NK)$ is Hurwitz, where K satisfies (6.70). With this Q, since $(A + B_2 NK)$ is Hurwitz, the Lyapunov equation (6.73) has a unique positive semidefinite solution R. Hence, to show that (6.72)–(6.75) admits a unique solution, it is sufficient to show that (6.74), (6.75) yields a unique solution (N, λ).

Recall that (6.74) can be substituted into (6.75) to yield (6.77). Furthermore, note that (6.78) can be rewritten as

$$h(N) = h_1(N)h_2(N), \qquad (8.241)$$

where

$$h_1(N) = \text{erf}^{-1}(N), \qquad (8.242)$$

$$h_2(N) = \left(N\sqrt{\pi}\exp\left(\text{erf}^{-1}(N)^2\right) - 2\text{erf}^{-1}(N)\right). \qquad (8.243)$$

Taking the derivatives of these functions with respect to N, one can show

$$h_1'(N), h_1''(N), h_2'(N), h_2''(N) > 0.$$

Moreover, $h_1(0) = h_2(0) = 0$ and, thus,

$$h'(N) > 0, \ h''(N) > 0. \qquad (8.244)$$

Clearly, $h(0) = 0$ and, furthermore, note that $h(N)$ has a vertical asymptote at $N = 1$. Thus, since $\rho, \eta > 0$, the left-hand side of (6.77) changes sign exactly once in the interval $N \in (0,1)$. Hence, (6.77) yields a unique solution N. Using this N, the value of λ is uniquely determined by either (6.74) or (6.75).

Since the optimization has an achievable solution, and the necessary conditions are uniquely satisfied, (6.70) and (6.71) constitute the globally minimizing solution. The cost (6.76) follows directly. $\qquad \square$

Proof of Theorem 6.8. First, we show that the partial derivatives of $\gamma^2(\rho, \eta)$ are nonnegative and then prove parts (i)–(iii) of the theorem.

Define $\tau(\rho, \eta)$ as

$$\tau(\rho, \eta) = \frac{\rho + \lambda(\rho, \eta)}{N^2(\rho, \eta)}. \qquad (8.245)$$

It is clear from (6.73) that $\gamma^2(\rho, \eta)$ depends on (ρ, η) through $\tau(\rho, \eta)$. Thus,

$$\frac{\partial}{\partial \rho} \gamma^2(\rho, \eta) = \frac{\partial}{\partial \rho} \operatorname{tr} \left\{ C_1 R(\tau(\rho, \eta)) C_1^T \right\}$$

$$= \frac{d}{d\tau} \operatorname{tr} \left\{ C_1 R(\tau) C_1^T \right\} \frac{\partial}{\partial \rho} \tau(\rho, \eta) \tag{8.246}$$

and, similarly,

$$\frac{\partial}{\partial \eta} \gamma^2(\rho, \eta) = \frac{d}{d\tau} \operatorname{tr} \left\{ C_1 R(\tau) C_1^T \right\} \frac{\partial}{\partial \eta} \tau(\rho, \eta). \tag{8.247}$$

To investigate $\operatorname{tr} \left\{ C_1 R'(\tau) C_1^T \right\}$, we substitute (8.245) into (6.72) and (6.73) to obtain

$$\left[A - \frac{1}{\tau} B_2 B_2^T Q(\tau) \right]^T Q(\tau) + Q(\tau) \left[A - \frac{1}{\tau} B_2 B_2^T Q(\tau) \right]$$

$$+ \frac{1}{\tau} Q(\tau) B_2 B_2^T Q(\tau) + C_1^T C_1 = 0, \tag{8.248}$$

$$\left[A - \frac{1}{\tau} B_2 B_2^T Q(\tau) \right] R(\tau) + R(\tau) \left[A - \frac{1}{\tau} B_2 B_2^T Q(\tau) \right]^T + B_1 B_1^T = 0. \tag{8.249}$$

Differentiating these equations with respect to τ results in

$$\left[A - \frac{1}{\tau} B_2 B_2^T Q(\tau) \right]^T Q'(\tau) + Q'(\tau) \left[A - \frac{1}{\tau} B_2 B_2^T Q(\tau) \right]$$

$$+ \frac{1}{\tau^2} Q(\tau) B_2 B_2^T Q(\tau) = 0, \tag{8.250}$$

$$\left[A - \frac{1}{\tau} B_2 B_2^T Q(\tau) \right] R'(\tau) + R'(\tau) \left[A - \frac{1}{\tau} B_2 B_2^T Q(\tau) \right]^T$$

$$+ \frac{1}{\tau} B_2 B_2^T \left[\frac{1}{\tau} Q(\tau) - Q'(\tau) \right] R(\tau) + \frac{1}{\tau} R(\tau) \left[\frac{1}{\tau} Q(\tau) - Q'(\tau) \right] B_2 B_2^T = 0. \tag{8.251}$$

To simplify the notation, we omit below the arguments of Q and R. Premultiplying (8.249) by Q and subtracting from (8.248), we obtain

$$\frac{1}{\tau} B_2^T Q R Q B_2 + \operatorname{tr} \left\{ C_1 R C_1^T \right\} - \operatorname{tr} \left\{ B_1^T Q B_1 \right\} = 0. \tag{8.252}$$

Taking the derivative of this equation with respect to τ gives

$$- \frac{1}{\tau^2} B_2^T Q R Q B_2 + \frac{1}{\tau} B_2^T \left[Q' R Q + Q R' Q + Q R Q' \right] B_2$$

$$+ \operatorname{tr} \left\{ C_1 R' C_1^T \right\} - \operatorname{tr} \left\{ B_1^T Q' B_1 \right\} = 0. \tag{8.253}$$

Premultiply (8.251) by Q, postmultiply (8.248) by R', and subtract the latter from the former. Taking the trace of the result yields

$$\operatorname{tr}\left\{C_1 R' C_1^T\right\} - \frac{2}{\tau^2} B_2^T QRQ B_2 + \frac{1}{\tau} B_2^T \left[Q'RQ + QR'Q + QRQ'\right] B_2 = 0. \quad (8.254)$$

From (8.253) and (8.254) we obtain

$$\frac{1}{\tau^2}\left\{B_2^T QRQ B_2\right\} = \operatorname{tr}\left\{B_1 Q' B_1^T\right\}. \quad (8.255)$$

Taking the derivative with respect to τ and using (8.254), we have

$$\operatorname{tr}\left\{C_1 R' C_1^T\right\} = -\tau \operatorname{tr}\left\{B_1 Q'' B_1^T\right\}. \quad (8.256)$$

Differentiating (8.250) with respect to τ results in

$$\left[A - \frac{1}{\tau} B_2 B_2^T Q\right]^T Q'' + Q''\left[A - \frac{1}{\tau} B_2 B_2^T Q\right] - \frac{2}{\tau}\left[\frac{1}{\tau} Q - Q'\right] B_2 B_2^T \left[\frac{1}{\tau} Q - Q'\right] = 0. \quad (8.257)$$

Since the matrix $(A - \frac{1}{\tau} B_2 B_2^T Q)$ is Hurwitz and the pair (A, B_2) is stabilizable, (8.254) implies that $Q''(\tau) \le 0$ and, thus, from (8.256)

$$\operatorname{tr}\left\{C_1 R'(\tau) C_1^T\right\} \ge 0. \quad (8.258)$$

Now, using (6.74), (6.75), we can differentiate τ with respect to η and ρ to obtain

$$\frac{\partial}{\partial \eta} \tau(\rho, \eta) = \frac{2 \operatorname{erf}^{-1}(N)^2}{N^2} > 0, \quad (8.259)$$

$$\frac{\partial}{\partial \rho} \tau(\rho, \eta) = \frac{1}{N^2} > 0. \quad (8.260)$$

Thus, based on (8.258)–(8.260), we conclude that the partial derivatives in (8.246), (8.247) are nonnegative. This conclusion is used to prove parts (i)–(iii) of the theorem.

Part (i): Note that $\gamma^2(\rho, \eta)$ is bounded from below, and hence, its limit as ρ tends to 0^+ exists. Similarly, $\tau(\rho, \eta) > 0$, and, thus, from (8.260), its limit as ρ tends to 0^+ also exists. Using (6.74) and (6.75) and a few algebraic manipulations, it is possible to show that

$$\lim_{\rho \to 0^+} \tau(\rho, \eta) = \lim_{N \to 0^+} \frac{\eta \sqrt{\pi} \operatorname{erf}^{-1}(N) \exp\left(\operatorname{erf}^{-1}(N)^2\right)}{N} = \frac{\pi \eta}{2}, \quad (8.261)$$

and, thus, with $R(\tau)$ and $Q(\tau)$ from (8.248) and (8.249),

$$\lim_{\rho \to 0^+} \gamma^2(\rho, \eta) = \operatorname{tr}\left\{C_1 R\left(\frac{\pi \eta}{2}\right) C_1^T\right\}. \quad (8.262)$$

Then (6.80), (6.81), and (6.82) follow immediately from (8.262), (8.248), and (8.249), respectively, using the notation

$$\bar{R}_\eta := R\left(\frac{\pi\eta}{2}\right), \quad \bar{Q}_\eta := Q\left(\frac{\pi\eta}{2}\right). \tag{8.263}$$

It follows from Theorem 6.7 that the solution of (6.81) and (6.82) exists and is unique, with the property that $\bar{R}_\eta \geq 0$, $\bar{Q}_\eta \geq 0$, which proves (i).

Part (ii): In a similar fashion to part (i), note that the limits of $\gamma^2(\rho,\eta)$ and $\tau^2(\rho,\eta)$ as η tends to 0^+ exist. From (6.77), as η tends to 0^+, N tends to 1. From (6.74), as N tends to 1, λ tends to 0. Thus,

$$\lim_{\eta\to 0^+} \tau(\rho,\eta) = \rho, \tag{8.264}$$

and hence,

$$\lim_{\eta\to 0^+} \gamma^2(\rho,\eta) = \text{tr}\left\{C_1 R(\rho) C_1^T\right\}, \tag{8.265}$$

which yields (6.83), using the notation $\gamma_{\rho 0}^2 := \text{tr}\left\{C_1 R(\rho) C_1^T\right\}$. Note from (6.72) and (6.73) that finding $R(\rho)$ is equivalent to solving the conventional LQR problem, which proves (ii).

Part (iii): From (8.259) and (8.260), $1/\tau(\rho,\eta)$ is monotonically decreasing in ρ and η, and bounded from below by 0. Using (6.74) and (6.75), it is possible to show that

$$\lim_{\eta\to\infty} \frac{1}{\tau(\rho,\eta)} = \lim_{\rho\to\infty} \frac{1}{\tau(\rho,\eta)} = 0. \tag{8.266}$$

Since A is Hurwitz, it follows from (6.72) and (6.73) that

$$\lim_{\eta\to\infty} \gamma^2(\rho,\eta) = \lim_{\rho\to\infty} \gamma^2(\rho,\eta) = \text{tr}\left\{C_1 R_{OL} C_1^T\right\}, \tag{8.267}$$

where $R_{OL} \geq 0$ is the solution of the Lyapunov equation

$$A R_{OL} + R_{OL} A + B_1 B_1^T = 0. \tag{8.268}$$

Thus, (6.84) follows immediately from (8.267) using the notation

$$\gamma_{OL}^2 := \text{tr}\left\{C_1 R_{OL} C_1^T\right\}.$$

From (8.268), R_{OL} is the open loop covariance matrix, and hence, γ_{OL}^2 is the open loop output variance.   □

Proof of Theorem 6.9. Let (N,Q,R,λ) be the unique solution of (6.72)–(6.75), and let K be obtained from (6.70). We begin by establishing that the matrix $(A + B_2 K)$ is Hurwitz. Note that Q satisfies the Ricatti equation (6.72). Using straightforward algebraic manipulations, we obtain

$$\left|1 - K(j\omega I - A)^{-1} B_2 N\right|^2 = 1 + \frac{1}{\tau}\left|C_1(j\omega I - A)^{-1} B_2\right|^2, \tag{8.269}$$

where $\tau = (\rho + \lambda)/N^2$. From (8.269), the Nyquist plot of $-K(j\omega I - A)^{-1}B_2N$ never enters the unit disk centered at the point $(-1,0)$ in the complex plane. Thus, $A + B_2N\kappa K$ is Hurwitz for all $\kappa > 1/2$. The result follows by setting $\kappa = 1/N$. Then, the statements of the theorem are proved as follows:

Part (i): Note that x_G is an equilibrium point of (6.85) if

$$Ax_G + B_2\text{sat}_\alpha(Kx_G) = 0. \tag{8.270}$$

Since $(A + B_2K)$ is nonsingular, this equation implies that

$$x_G + (A + B_2K)^{-1}B_2[\text{sat}_\alpha(Kx_G) - Kx_G] = 0. \tag{8.271}$$

Premultiplying (8.271) by K yields

$$[1 - \Gamma]Kx_G + \Gamma\text{sat}_\alpha(Kx_G) = 0, \tag{8.272}$$

where

$$\Gamma = K(A + B_2K)^{-1}B_2. \tag{8.273}$$

Using the Schur complement, we have

$$1 - \Gamma = \det[I - (A + B_2K)^{-1}B_2K]$$
$$= \det[(A + B_2K)]\det(A) \geq 0. \tag{8.274}$$

It follows that $\Gamma \leq 1$, and hence, (8.272) is satisfied only if $Kx_G = 0$. Thus, from (8.271), $x_G = 0$ is the unique equilibrium.

Part (ii): The Jacobian linearization of (6.85) about $x_G = 0$ is given by

$$\Delta\dot{x}_G = (A + B_2K)\Delta x_G. \tag{8.275}$$

Since $(A + B_2K)$ is Hurwitz, the result follows from Lyapunov's indirect method.

Part (iii): To prove (iii), we establish asymptotic stability of the origin via Lyapunov function. Recall that (A, B_2) is stabilizable and (C_1, A) is detectable. Also, note that the origin of (6.85) is asymptotically stable if and only if it is asymptotically stable for the controllable and observable portion of its Kalman canonical decomposition. Thus, assume without loss of generality that (A, B_2) is controllable and (C_1, A) is observable. Then it follows from (6.72) that $Q > 0$. Consider the candidate Lyapunov function

$$V(x_G) = x_G^T(\varepsilon Q)x_G, \tag{8.276}$$

where

$$\varepsilon = \frac{N^2}{\rho + \lambda}. \tag{8.277}$$

It is straightforward to show that

$$\dot{V}(x_G) = -x_G^T(\varepsilon C_1^T C_1)x_G - \varepsilon(\rho + \lambda)[\frac{2u}{N}\text{sat}_\alpha(u) - u^2], \tag{8.278}$$

and thus,

$$\dot{V}(x_G) \leq 0 \tag{8.279}$$

if

$$|u| \leq 2\alpha/N. \tag{8.280}$$

Clearly, this is equivalent to

$$|B_2^T(\varepsilon Q)x_G| \leq 2\alpha, \tag{8.281}$$

and hence, $\dot{V}(x_G) \leq 0$ for all $x_G \in \mathcal{X}$, where

$$\mathcal{X} = \left\{ x_G \in R^{n_x} : x_G^T \left(Q B_2 B_2^T Q \right) x_G \leq \frac{4\alpha^2}{\varepsilon^2} \right\}. \tag{8.282}$$

Now, define the set $\mathcal{S}$ such that

$$\mathcal{S} = \left\{ x_G \in \mathcal{X} : \dot{V}(x_G) = 0 \right\}, \tag{8.283}$$

and assume that the trajectory $x_G(t)$ belongs to $\mathcal{S}$ for all t. Then, from (8.278), $C_1 x_G = 0$ and since (C_1, A) is observable, $x_G = 0$. Thus, by LaSalle's Theorem, the equilibrium point $x_G = 0$ is asymptotically stable and $\mathcal{X}$ is a subset of its domain of attraction. □

Proof of Theorem 6.10. The proof is similar to that of Theorem 6.7. First, we use the method of Lagrange multipliers to find the necessary conditions for optimality. To verify the regularity of the contraints, let $(K, L, M, N_a, N_s, \tilde{P}, \alpha, \beta)$ satisfy (6.96) and (6.97), and for arbitrary symmetric matrices $\tilde{Q}$ and $\Lambda = \mathrm{diag}(\lambda_1, \lambda_2)$, define

$$\Phi\left(K, L, M, N_a, N_s, \tilde{P}, \alpha, \beta\right) = \mathrm{tr}\left\{ \left[\left(\tilde{A} + \tilde{B}_2 \tilde{N} \tilde{C}_2 \right) \tilde{P} + \tilde{P}\left(\tilde{A} + \tilde{B}_2 \tilde{N} \tilde{C}_2 \right)^T + \tilde{B}_1 \tilde{B}_1^T \right] \tilde{Q} \right\}$$
$$+ \mathrm{tr}\left\{ \Lambda \left[\mathrm{diag}\left\{ \tilde{C}_2 \tilde{P} \tilde{C}_2^T \right\} - \frac{1}{2}\Theta\left[\mathrm{erf}^{-1}\left(\tilde{N} \right) \right]^{-2} \right] \right\}. \tag{8.284}$$

Differentiating Φ with respect to α, β, and $\tilde{P}$, and the setting the result to zero yields

$$\frac{-2\lambda_1 \alpha}{2\mathrm{erf}^{-1}(N_a)^2} = 0, \tag{8.285}$$

$$\frac{-2\lambda_2 \beta}{2\mathrm{erf}^{-1}(N_s)^2} = 0, \tag{8.286}$$

$$\left(\tilde{A} + \tilde{B}_2 \tilde{N} \tilde{C}_2 \right)^T \tilde{Q} + \tilde{Q}\left(\tilde{A} + \tilde{B}_2 \tilde{N} \tilde{C}_2 \right) + \Lambda \tilde{C}_2^T \tilde{C}_2 = 0. \tag{8.287}$$

Since $N_a \in (0,1)$, $N_s \in (0,1)$, and $\alpha, \beta > 0$, it follows from (8.285) and (8.286) that $\lambda_1 = \lambda_2 = 0$, which, from (8.287), implies that $\tilde{Q} = 0$. It is straightforward to show that, consequently,

$$\frac{\partial \Phi}{\partial K} = \frac{\partial \Phi}{\partial L} = \frac{\partial \Phi}{\partial M} = \frac{\partial \Phi}{\partial N_a} = \frac{\partial \Phi}{\partial N_s} = 0, \tag{8.288}$$

and, thus, the constraints are regular.

Consider the Lagrangian

$$\Psi\left(K, L, M, N_a, N_s, \tilde{P}, \tilde{Q}, \alpha, \beta, \lambda_1, \lambda_2\right) = \text{tr}\left\{\tilde{C}_1 \tilde{P} \tilde{C}_1^T\right\}$$

$$+ \text{tr}\left\{\left[\left(\tilde{A} + \tilde{B}_2 \tilde{N} \tilde{C}_2\right)\tilde{P} + \tilde{P}\left(\tilde{A} + \tilde{B}_2 \tilde{N} \tilde{C}_2\right)^T + \tilde{B}_1 \tilde{B}_1^T\right]\tilde{Q}\right\}$$

$$+ \text{tr}\left\{\Lambda\left[\text{diag}\left\{\tilde{C}_2 \tilde{P} \tilde{C}_2^T\right\} - \frac{1}{2}\Theta\left[\text{erf}^{-1}\left(\tilde{N}\right)\right]^{-2}\right]\right\}. \tag{8.289}$$

and the partition

$$\tilde{P} = \begin{bmatrix} P_{11} & P_{12}^T \\ P_{12} & P_{22} \end{bmatrix}, \tilde{Q} = \begin{bmatrix} Q_{11} & Q_{12} \\ Q_{12}^T & Q_{22} \end{bmatrix}.$$

Differentiating Ψ with respect to K, L, M, N_a, N_s, $\tilde{P}$, $\tilde{Q}$, λ_1, λ_2, α, β, and equating the results to zero yields the necessary conditions for optimality

$$KP_{22} + \frac{N_a}{\rho + \lambda_1}\left(Q_{11}P_{12}^T + Q_{12}P_{22}\right) = 0, \tag{8.290}$$

$$Q_{22}L - \left(Q_{12}^T P_{11} + Q_{22}P_{12}\right)C_2^T \frac{N_s}{\mu} = 0, \tag{8.291}$$

$$Q_{12}^T P_{12}^T + Q_{22}P_{22} = 0, \tag{8.292}$$

$$K\left(P_{12}Q_{11} + P_{22}Q_{12}^T\right)B_2 + \frac{\sqrt{\pi}}{4}\lambda_1 \exp\left(\text{erf}^{-1}(N_a)^2\right) \times \left(\text{erf}^{-1}(N_a)^{-3}\right), \tag{8.293}$$

$$C_2\left(P_{11}Q_{12} + P_{12}^T Q_{22}\right)L + \frac{\sqrt{\pi}}{4}\lambda_2 \exp\left(\text{erf}^{-1}(N_s)^2\right) \times \left(\text{erf}^{-1}(N_s)^{-3}\right) = 0, \tag{8.294}$$

$$\left(\tilde{A} + \tilde{B}_2 \tilde{N} \tilde{C}_2\right)^T \tilde{Q} + \tilde{Q}\left(\tilde{A} + \tilde{B}_2 \tilde{N} \tilde{C}_2\right) + \tilde{C}_1^T \tilde{C}_1 + \tilde{C}_2^T \Lambda \tilde{C}_2 = 0, \tag{8.295}$$

$$\left(\tilde{A} + \tilde{B}_2 \tilde{N} \tilde{C}_2\right)\tilde{P} + \tilde{P}\left(\tilde{A} + \tilde{B}_2 \tilde{N} \tilde{C}_2\right)^T + \tilde{B}_1 \tilde{B}_1^T = 0, \tag{8.296}$$

$$KP_{22}K^T - \frac{\alpha^2}{2\text{erf}^{-1}(N_a)^2} = 0, \tag{8.297}$$

$$C_2 P_{11} C_2^T - \frac{\beta^2}{2\text{erf}^{-1}(N_s)^2} = 0, \tag{8.298}$$

$$2\eta_a \alpha - \frac{2\lambda_1 \alpha}{2\text{erf}^{-1}(N_a)^2} = 0, \tag{8.299}$$

$$2\eta_s \alpha - \frac{2\lambda_2 \beta}{2\text{erf}^{-1}(N_s)^2} = 0. \tag{8.300}$$

In the subsequent analysis we assume that P_{22} and Q_{22} are invertible, although similar results can be obtained by using pseudoinverses. From (8.290) and (8.291), we obtain

$$K = -\frac{N_a}{\rho + \lambda_1} B_2^T \left(Q_{11} P_{12}^T + Q_{12} P_{22} \right) P_{22}^{-1}, \qquad (8.301)$$

$$L = Q_{22}^{-1} \left(Q_{12}^T P_{11} + Q_{22} P_{12} \right) C_2^T \frac{N_s}{\mu}. \qquad (8.302)$$

Defining $T := P_{12}^T P_{22}^{-1}$, it follows from (8.301) that

$$K = -\frac{N_a}{\rho + \lambda_1} B_2^T \left(Q_{11} - Q_{12} Q_{22}^{-1} Q_{12}^T \right) T. \qquad (8.303)$$

Note from (8.292) that $T^{-1} = -Q_{22}^{-1} Q_{12}^T$, and, thus, from (8.302)

$$L = -T^{-1} \left(P_{11} + P_{12}^T P_{22}^{-1} P_{12} \right) C_2^T \frac{N_s}{\mu}. \qquad (8.304)$$

Then, with

$$Q = Q_{11} - Q_{12} Q_{22}^T Q_{12}^T, \quad P = P_{11} + P_{12}^T P_{22}^{-1} P_{12}, \qquad (8.305)$$

(8.303) and (8.304) become

$$K = -\frac{N_a}{\rho + \lambda_1} B_2^T Q T, \qquad (8.306)$$

and

$$L = -\frac{N_a}{\mu} T^{-1} P C_2^T. \qquad (8.307)$$

By substituting (8.306) and (8.307) into (8.295) and (8.296), it is straightforward to show that

$$M = T^{-1} \left(A + B_2 N_a K T^{-1} + T L N_s C_2 \right) T. \qquad (8.308)$$

The equations (6.99)–(6.101) for the parameters K, L, M follow immediately from (8.306)–(8.308), noting that T is a similarity transformation and does not affect the the controller transfer function $K(sI - M)^{-1} L$. The equations (6.102) and (6.103) for α and β follow directly from (8.297) and (8.298), respectively.

We now verify (6.104)–(6.111). Equations (6.104)–(6.107) follow by substituting (8.306)–(8.308) into (8.295) and (8.296), where

$$R = P_{12}^T P_{22}^{-1} P_{12}, \; S = Q_{12} Q_{22}^{-1} Q_{12}^T. \tag{8.309}$$

Multiplying (8.290) from the right by K^T, substituting into (8.293), and then using (8.297), yields (6.108). Multiplying (8.291) from the left by L^T and substituting into (8.294) yields (6.109). Finally, (6.110) and (6.111) follow immediately from (8.299) and (8.300).

We now argue that the ILQG problem (6.94) has a solution, that is, the minimum exists and is attained. Note that (6.94) can be reformulated as

$$\min_{\alpha, \beta} \left\{ \eta_a \alpha^2 + \eta_s \beta^2 + \min_{K,L,M} \sigma_{\hat{z}}^2 \right\}. \tag{8.310}$$

It is known from SLQG theory (Section 6.1) that, under Assumption 6.1, for every $\alpha, \beta > 0$, the triple (K, L, M) that solves

$$\min_{K,L,M} \sigma_{\hat{z}}^2, \tag{8.311}$$

exists. Moreover, the achieved minimum is continuous in α and β. Thus, the function

$$q_G(\alpha, \beta) = \eta_a \alpha^2 + \eta_s \beta^2 + \min_{K,L,M} \sigma_{\hat{z}}^2 \tag{8.312}$$

is continuous for $\alpha, \beta > 0$ and tends to ∞ as $\alpha \to \infty$ and $\beta \to \infty$. Hence, $q_G(\alpha, \beta)$ achieves a minimum at some $\alpha^* \geq 0$, $\beta^* \geq 0$. Let (K^*, L^*, M^*) be the minimizer of (8.311) when $\alpha = \alpha^*$ and $\beta = \beta^*$. Then the quintuple $(K^*, L^*, M^*, \alpha^*, \beta^*)$ solves (6.94).

The ILQG cost (6.112) follows immediately from (6.99)–(6.103). Since the optimization has an achievable solution, any solution of (6.104)–(6.111) that minimizes (6.112) solves the ILQG problem. □

Proof of Theorem 6.11. Let $(N_a, N_s, \lambda_1, \lambda_2, P, Q, R, S)$ be the minimizing solution of (6.104)–(6.111) and let K and L be obtained from (6.99) and (6.100). Similar to the proof of Theorem 6.9, we first establish that $(A + B_2 K)$ and $(A + LC_2)$ are Hurwitz. Since P and Q satisfy (6.104) and (6.105), it is readily shown that

$$\left| 1 - K (j\omega I - A)^{-1} B_2 N_a \right|^2 = 1 + \frac{1}{\tau_1} \left| C_1 (j\omega I - A)^{-1} B_2 \right|^2 + \frac{\lambda_2}{\tau_1} \left| C_2 (j\omega I - A)^{-1} B_2 \right|^2, \tag{8.313}$$

and

$$\left| 1 - C_2 (j\omega I - A)^{-1} N_s L \right|^2 = 1 + \frac{1}{\tau_2} \left| C_2 (j\omega I - A)^{-1} B_1 \right|^2, \tag{8.314}$$

where $\tau_1 = (\rho + \lambda_1)/N_a^2$ and $\tau_2 = \mu/N_s^2$. Thus, $(A + B_2 N_a \kappa_a K)$ and $(A + LN_s \kappa_s C_2)$ are Hurwitz for all $\kappa_a > 1/2$, $\kappa_s > 1/2$, and the result follows by setting $\kappa_a = 1/N_a$ and $\kappa_s = 1/N_s$.

Part (i): The point $[x_G, x_C]$ is an equilibrium of the system if

$$Ax_G + B_2 \text{sat}_\alpha (Kx_C) = 0, \tag{8.315}$$

$$Ax_C + B_2 \text{sat}_\alpha (Kx_C) - L \left(y - \text{sat}_\beta (C_2 x_C) \right) = 0, \tag{8.316}$$

or, equivalently,

$$Ax_G + B_2 \text{sat}_\alpha (Kx_C) = 0, \tag{8.317}$$

$$Ae + L \left(\text{sat}_\beta (C_2 x_G) - \text{sat}_\beta (C_2 x_C) \right) = 0, \tag{8.318}$$

where $e = x_G - x_C$. Since $(A + B_2 K)$ and $(A + LC_2)$ are nonsingular, we can write

$$x_G + (A + B_2 K)^{-1} B_2 \text{sat}_\alpha (Kx_C) - (A + B_2 K)^{-1} B_2 Kx_G = 0, \tag{8.319}$$

$$e + (A + LC_2)^{-1} L \left(\text{sat}_\beta (C_2 x_G) - \text{sat}_\beta (C_2 x_C) \right) - (A + LC_2)^{-1} LC_2 e = 0. \tag{8.320}$$

Premultiplying (8.320) by C_2 results in

$$(1 - \Gamma_C) C_2 x_G + \Gamma_C \text{sat}_\beta (C_2 x_G) = (1 - \Gamma_C) C_2 x_C + \Gamma_C \text{sat}_\beta (C_2 x_C), \tag{8.321}$$

where $\Gamma_C = C_2 (A + LC_2)^{-1} L$. Since

$$1 - \Gamma_C = \det (A + LC_2)^{-1} \det (A) \geq 0, \tag{8.322}$$

it follows that $\Gamma_C \leq 1$. If $\Gamma_C \neq 1$, then (8.321) implies that $C_2 x_G = C_2 x_C$, which, from (8.320), implies $e = 0$, that is, $x_G = x_C$. Then, it follows from (8.315), (8.319) and the proof of Theorem 6.9 that $x_G = 0$, and, necessarily, $x_C = 0$. If $\Gamma_C = 1$, then from (8.321)

$$\text{sat}_\beta (C_2 x_G) = \text{sat}_\beta (C_2 x_C), \tag{8.323}$$

which, from (8.318) implies that $Ae = 0$. Then, from (8.315)

$$Ax_C + B_2 \text{sat}_\alpha (Kx_C) = 0, \tag{8.324}$$

which, through the proof of Theorem 6.9, implies that $x_C = 0$. Thus, from (8.323), $C_2 x_G = 0$, and it follows from (8.320) that $e = 0$. Thus, $x_G = 0$ and the result is established.

Part (ii): The Jacobian linearization of the system about the equilibrium $[x_G, x_C] = 0$ is given by

$$\Delta \dot{x}_G = (A + B_2 K) \Delta x_G, \tag{8.325}$$

$$\Delta \dot{e} = (A + LC_2) \Delta e, \tag{8.326}$$

Since $(A + B_2 K)$ and $(A + LC_2)$ are Hurwitz, the result follows from Lyapunov's indirect method.

Part (iii): As in the proof of Theorem 6.9, assume without loss of generality that (A, B_2) is controllable and (C_1, A) is observable. Consider the candidate Lyapunov function

$$V(x_G, e) = V_1(x_G) + V_2(e), \tag{8.327}$$

where

$$V_1(x_G) = x_G^T (\varepsilon_1 Q) x_G, \tag{8.328}$$

$$V_2(e) = e^T M e, \tag{8.329}$$

where $\varepsilon_1 = N_a^2/(\rho + \lambda_1)$, and where, since $(A + LC_2)$ is Hurwitz, M is the positive definite solution of

$$(A + LC_2)^T M + M(A + LC_2) + I = 0. \tag{8.330}$$

It follows that

$$\dot{V}_1(x_G) = -x_G^T(\varepsilon_1 C_1^T C_1 + \varepsilon_1 \lambda_2 C_2^T C_2) x_G - \varepsilon_1(\rho + \lambda_1)[\frac{2u}{N}\mathrm{sat}_\alpha(u) - u^2], \tag{8.331}$$

$$\dot{V}_2(e) = \left(e^T A^T + \left(\mathrm{sat}_\beta(C_2 x_G) - \mathrm{sat}_\beta(C_2 x_C)\right)^T L^T\right) M e$$
$$+ e^T M \left(A e + L \left(\mathrm{sat}_\beta(C_2 x_G) - \mathrm{sat}_\beta(C_2 x_C)\right)\right). \tag{8.332}$$

Thus, $\dot{V}_1(x_G) \le 0$ if

$$|u| \le \frac{2\alpha}{N}, \tag{8.333}$$

and by the proof of Theorem 6.9, $\dot{V}_1(x_G) \le 0$ for all $x_G \in \mathcal{X}_1$, where

$$\mathcal{X}_1 = \left\{ x_G \in R^{n_x} | x_G^T \left(Q B_2 B_2^T Q\right) x_G \le \frac{4\alpha^2}{\varepsilon_1^2} \right\} \tag{8.334}$$

Now, assume that $x_G = x_C$. It follows from (8.332) and (8.330) that

$$\dot{V}_2(e) = -e^T e, \tag{8.335}$$

if

$$|C_2 x_G| \le \beta, \tag{8.336}$$

and thus, $\dot{V}_2(e) \le 0$ for all $x_G \in \mathcal{X}_2$, where

$$\mathcal{X}_2 = \left\{ x_G \in R^{n_x} | x_G^T C_2^T C_2 x_G \le \beta^2 \right\}. \tag{8.337}$$

Then, from (8.334) and (8.337), $\dot{V}(x_G, x_C) \le 0$ for all $(x_G, x_C) \in \mathcal{Y} \times \mathcal{Y}$, where

$$\mathcal{Y} = \mathcal{X}_1 \cap \mathcal{X}_2. \tag{8.338}$$

Now, define the set S such that

$$S = \{(x_G, x_C) \in \mathcal{Y} \times \mathcal{Y} \mid \dot{V}(x_G, x_C) = 0\}, \tag{8.339}$$

and assume that the trajectory $(x_G(t), x_C(t))$ belongs to S for all t. Then it follows from (8.331)–(8.335) that $C_1 x_G = 0$ and $e = 0$. Thus, since (C_1, A) is observable, $x_G = x_C = 0$, and by LaSalle's theorem, $[x_G, x_C] = 0$ is asymptotically stable and $\mathcal{Y} \times \mathcal{Y}$ is a subset of its domain of attraction. $\square$

8.6 Proofs for Chapter 7

To prove Theorem 7.1, we need the following

Lemma 8.6. *Let statements (i) and (ii) of Assumption 7.1 hold and r be such that the Nyquist plot of the loop gain $L = PC$ lies entirely outside of $D(r)$. Then*

$$\left\| \frac{FP}{1 + NL} \right\|_2 < \frac{2r}{(2r+1)N - 1} \sigma_{z_l}, \quad \forall N \in (\frac{1}{2r+1}, 1],$$

$$\left\| \frac{FPC}{1 + NL} \right\|_2 < \frac{2r}{(2r+1)N - 1} \sigma_{u_l}, \quad \forall N \in (\frac{1}{2r+1}, 1].$$

Proof. First, we prove the existence of $\|\frac{FP}{1+NL}\|_2$, $\|\frac{FPC}{1+NL}\|_2$, and $\|\frac{1+L}{1+NL}\|_\infty$ for all $N \in (\frac{1}{2r+1}, 1]$. The first two norms exist if $\frac{FP}{1+NL}$ and $\frac{FPC}{1+NL}$ are asymptotically stable and strictly proper for all $N \in (\frac{1}{2r+1}, 1]$. The asymptotic stability follows from the fact that $\frac{FP}{1+L}$ and $\frac{FPC}{1+L}$ are asymptotically stable (due to Assumption 7.1 (i)) and the number of encirclement of $-1 + j0$ by $L(j\omega)$ and $NL(j\omega)$ are the same (due to the fact that $L(j\omega)$ is outside of $D(r)$ and $\frac{1}{2r+1} < N \leq 1$). The strict properness of these two transfer functions follows from Assumption 7.1 (ii). Thus, $\|\frac{FP}{1+NL}\|_2$ and $\|\frac{FPC}{1+NL}\|_2$ exist for all $N \in (\frac{1}{2r+1}, 1]$.

Since, as it follows from the above, $\frac{1+L}{1+NL}$ is also asymptotically stable and, obviously, proper, $\|\frac{1+L}{1+NL}\|_\infty$ exists for all $N \in (\frac{1}{2r+1}, 1]$.

Next, we show that

$$\left\| \frac{FP}{1 + NL} \right\|_2^2 \leq \left\| \frac{1 + L}{1 + NL} \right\|_\infty \sigma_{y_l}, \quad \forall N \in (\frac{1}{2r+1}, 1], \tag{8.340}$$

$$\left\| \frac{FPC}{1 + NL} \right\|_2^2 \leq \left\| \frac{1 + L}{1 + NL} \right\|_\infty \sigma_{u_l}, \quad \forall N \in (\frac{1}{2r+1}, 1]. \tag{8.341}$$

Indeed, (8.340) can be obtained as follows

$$\left\| \frac{FP}{1 + NL} \right\|_2^2 = \frac{1}{2\pi} \int_{-\infty}^{\infty} \left| \frac{FP}{1 + NL} \right|^2 d\omega$$

$$= \frac{1}{2\pi} \int_{-\infty}^{\infty} \left| \frac{FP}{1 + L} \right|^2 \left| \frac{1 + L}{1 + NL} \right|^2 d\omega$$

$$\leq \frac{1}{2\pi} \left\| \frac{1+L}{1+NL} \right\|_\infty^2 \int_{-\infty}^{\infty} \left| \frac{FP}{1+L} \right|^2 d\omega$$

$$= \left\| \frac{1+L}{1+NL} \right\|_\infty^2 \sigma_{y_l}^2. \tag{8.342}$$

Inequality (8.341) is proved similarly.

In turn, $\|\frac{1+L}{1+NL}\|_\infty$, for each $N \in (\frac{1}{2r+1}, 1]$, is bounded as follow:

$$\left\| \frac{1+L}{1+NL} \right\|_\infty = \max_\omega \left| \frac{1+\mathrm{Re}L(j\omega)+j\mathrm{Im}L(j\omega)}{1+N\mathrm{Re}L(j\omega)+jN\mathrm{Im}L(j\omega)} \right|$$

$$\leq \max_{x+jy \notin D(r)} \left| \frac{1+x+jy}{1+Nx+jNy} \right|. \tag{8.343}$$

For each positive $c \neq \frac{1}{N}$, the level set $\left\{ (x,y) : |\frac{1+x+jy}{1+Nx+jNy}| = c \right\}$ is a circle given by

$$\left(x+1+\frac{c^2 N(N-1)}{1-c^2 N^2} \right)^2 + y^2 = \left(\frac{c(1-N)}{1-c^2 N^2} \right)^2. \tag{8.344}$$

It is possible to show that the level set $\left\{ (x,y) : |\frac{1+x+jy}{1+Nx+jNy}| = c \right\}$ is not entirely contained in $D(r)$ for all c small enough; that there exists a unique $c = c^*$ such that the circle (8.344) is tangent to and contained in $D(r)$; and that for all $c > c^*$, the level set is strictly contained in $D(r)$. This c^* is an upper-bound for the right-hand side of (8.343), that is,

$$\max_{x+jy \notin D(r)} \left| \frac{1+x+jy}{1+Nx+jNy} \right| < c^*. \tag{8.345}$$

The condition of tangency is:

$$(2r+1)-1-\frac{c^2 N(N-1)}{1-c^2 N^2} = \frac{c(N-1)}{1-c^2 N^2}, \tag{8.346}$$

that is,

$$c^* = \frac{2r}{(2r+1)N-1}, \quad \forall N \in (\frac{1}{2r+1}, 1]. \tag{8.347}$$

Therefore,

$$\left\| \frac{1+L}{1+NL} \right\|_\infty \leq \frac{2r}{(2r+1)N-1}, \quad \forall N \in (\frac{1}{2r+1}, 1], \tag{8.348}$$

which, along with (8.340) and (8.341), completes the proof.

Proof of Theorem 7.1. First, we show that $\alpha \geq \beta(e,r)\sigma_{u_l}$ implies

$$\left\| \frac{FPC}{1+N_1 L} \right\|_2 \leq \frac{\alpha}{\sqrt{2}\,\mathrm{erf}^{-1}(N_1)}, \tag{8.349}$$

where

$$N_1 = \frac{2r+(1+e)}{(1+e)(2r+1)} > \frac{1}{2r+1}. \tag{8.350}$$

Indeed, rewriting $\alpha \geq \beta(e,r)\sigma_{u_l}$ using (8.350) and (7.9) yields

$$\alpha \geq \beta(e,r)\sigma_{u_l}$$
$$= \sqrt{2}(1+e)\mathrm{erf}^{-1}\left(\frac{2r+(1+e)}{(1+e)(2r+1)}\right)\sigma_{u_l}$$
$$= \sqrt{2}(1+e)\mathrm{erf}^{-1}(N_1)\sigma_{u_l}. \tag{8.351}$$

Then, from Lemma 8.6, (8.350) and (8.351), we obtain

$$\left\|\frac{FPC}{1+N_1L}\right\|_2 \leq \frac{2r}{(2r+1)N_1-1}\sigma_{u_l} = (1+e)\sigma_{u_l} \leq \frac{\alpha}{\sqrt{2}\,\mathrm{erf}^{-1}(N_1)}.$$

Next, we show that the quasilinear gain N of the system of Figure 7.1(c) corresponding to a given $\alpha > \beta(e,r)\sigma_{u_l}$ exists and, moreover, $N \geq N_1$. Define a function

$$h(x) = \left\|\frac{FPC}{1+xL}\right\|_2 - \frac{\alpha}{\sqrt{2}\,\mathrm{erf}^{-1}(x)}. \tag{8.352}$$

Then, as the gain N is defined by

$$\left\|\frac{FPC}{1+NL}\right\|_2 = \frac{\alpha}{\sqrt{2}\,\mathrm{erf}^{-1}(N)}, \tag{8.353}$$

$x = N$ solves $h(x) = 0$. For $x = 1$, due to the fact that $\mathrm{erf}^{-1}(1) = \infty$, we can write:

$$h(1) = \left\|\frac{FPC}{1+L}\right\|_2 \geq 0. \tag{8.354}$$

From (8.349), we obtain

$$h(N_1) \leq 0. \tag{8.355}$$

Since, $h(x)$ is continuous in x, from (8.354) and (8.355) we conclude that there exists $N \in [N_1, 1]$ satisfying (8.353).

Finally, using Lemma 8.6, we show that $\sigma_y \leq (1+e)\sigma_{y_l}$:

$$\sigma_y = \left\|\frac{FP}{1+NL}\right\|_2 \leq \frac{2r}{(2r+1)N-1}\sigma_{y_l}$$
$$\leq \frac{2r}{(2r+1)N_1-1}\sigma_{y_l}$$
$$= (1+e)\,\sigma_{y_l}. \tag{8.356}$$

This completes the proof. $\square$

Proof of Theorem 7.2. *Sufficiency:* From (7.24) and (7.25),

$$K_a\mathcal{F}\left(K_a\left\|\frac{P(s)C(s)}{1+P(s)C(s)}\right\|_2\right) = 1. \tag{8.357}$$

It follows from (8.357) that (7.19) is a solution of (7.23). Thus, a-boosting is possible with the boosting gain K_a.

Necessity: a-Boosting is possible with the boosting gain K_a. Thus, (7.19) and (7.23) hold. Clearly, substituting the former into the latter yields (7.24) and (7.25). □

Proof of Theorem 7.3. It is straightforward to show that the function

$$h(x) = x \text{erf}\left(\frac{c}{x}\right) \tag{8.358}$$

is continuous and monotonically increasing $\forall c > 0$, with the property that

$$h(0) = 0 \tag{8.359}$$

and

$$\lim_{x \to \infty} h(x) = \frac{2}{\sqrt{\pi}} c. \tag{8.360}$$

Hence, (7.28) admits a positive solution if and only if

$$\frac{2}{\sqrt{\pi}} c > 1, \tag{8.361}$$

which is equivalent to (7.30). Moreover, any positive solution of (7.28) must be unique because $h(x)$ defined in (8.358) is monotonically increasing. □

Proof of Theorem 7.4. Observe from Figure 7.7(c) that

$$K_a N_a = K_a \mathcal{F}\left(\left\|\frac{P(s) C(s) N_s K_s K_a}{1 + P(s) N_s K_s N_a K_a C(s)}\right\|_2\right) \tag{8.362}$$

and

$$K_s N_s = K_s \mathcal{G}\left(\left\|\frac{P(s)}{1 + P(s) N_s K_s N_a K_a C(s)}\right\|_2\right). \tag{8.363}$$

Substituting

$$K_a N_a = K_s N_s = 1 \tag{8.364}$$

into (8.362) and (8.363) yields (7.23) and (7.35), which establishes the separation principle. □

8.7 Annotated Bibliography

The material of Section 8.1 is based on the following:

[8.1] C. Gokcek, "Disturbance Rejection and Reference Tracking in Control Systems with Saturating Actuators, Ph. D. Dissertation, Department of Electrical Engineering and Computer Science, University of Michigan, Ann Arbor, MI, 2000

The material of Section 8.2 is based on the following:

[8.2] Y.Eun, P.T. Kabamba, and S.M. Meerkov, "System types in feedback control with saturating actuators," *IEEE Transactions on Automatic Control*, Vol. 49, pp. 287–291, 2004

[8.3] Y.Eun, P.T. Kabamba, and S.M. Meerkov, "Tracking random references: Random sensitivity function and tracking quality indicators," *IEEE Transactions on Automatic Control*, Vol. 48, pp. 1666–1671, 2003

[8.4] Y.Eun, P.T. Kabamba, and S.M. Meerkov, "Analysis of random reference racking in systems with saturating actuators," *IEEE Transactions on Automatic Control*, Vol. 50, pp. 1861–1866, 2005

[8.5] Y. Eun, Reference Tracking in Feedback Control with Saturating Actuators, Ph. D. Dissertation, Department of Electrical Engineering and Computer Science, University of Michigan, Ann Arbor, MI, 2003

The material of Section 8.3 is based on the reference. The material of Section 8.4 is based on the following:

[8.6] S. Ching, P.T. Kabamba, and S.M. Meerkov, "Admissible pole locations for tracking random references," *IEEE Transactions on Automatic Control*, Vol. 54, pp. 168–171, 2009

[8.7] S. Ching, P.T. Kabamba, and S.M. Meerkov, "Root locus for random reference tracking in systems with saturating actuators," *IEEE Transactions on Automatic Control*, Vol. 54, pp. 79–91, 2009

[8.8] S. Ching, Control methods for Systems with Nonlinear Instrumentation: Roof Locus, Performance Recovery, and Instrumented LQG, Ph. D. Dissertation, Department of Electrical Engineering and Computer Science, University of Michigan, Ann Arbor, MI, 2009

The material of Section 8.5 is based on reference [8.1], [8.8], and the following:

[8.9] C. Gokcek, P.T. Kabamba, and S.M. Meerkov, "An LQR/LQG theory for systems with saturating actuators," *IEEE Transactions on Automatic Control*, Vol. 46, No. 10, pp. 1529–1542, 2001

[8.10] S. Ching, P.T. Kabamba, and S.M. Meerkov, "Instrumented LQR/LQG: A method for simultaneous design of controller and instrumentation," submitted to IEEE Transactions on Automatic Control, Vol. 55–1, pp. 217–221.

The material of Section 8.6 is based on reference [8.5], [8.8], and the following:

[8.11] Y. Eun, C. Gokcek, P.T. Kabamba, and S.M. Meerkov, "Selecting the level of actuator saturation for small performance degradation of linear designs," in *Actuator Saturation Control*, V. Kapila and K.M. Grigiriadis, Eds., Marcel Dekker, Inc., New York, pp. 33–45, 2002

[8.12] S. Ching, P.T. Kabamba, and S.M. Meerkov, "Recovery of linear performance in feedback systems with nonlinear instrumentation," *Proc of the 2009 American Control Conference*, Vols. 1–9, pp. 2545–2550, 2009

Epilogue

This volume has extended major ideas and results of linear control theory to systems with nonlinear actuators and sensors. Namely, the linear frequency domain methods (based on the frequency of harmonic inputs) are extended to the quasilinear frequency domain (based on the bandwidth of random inputs). The linear time domain methods are extended to quasilinear ones by introducing tracking quality indicators (instead of overshoot and settling time) and by modifying the root locus technique (e.g., the saturated root locus). The LQR/LQG methodology has been extended to SLQR/SLQG (where the "S" stands for "saturated"). Although not included in this volume, the H_∞ and LMI techniques have also been extended to systems with nonlinearities in sensors and actuators.

In addition, new problems, specific for LPNI systems, have been formulated and solved. These include Instrumented LQR/LQG, where the optimal controller and instrumentation are synthesized simultaneously, and the problem of performance recovery, where the performance losses due nonlinear instrumentation are either contained to a desired level or eliminated altogether.

What made these results possible?

1. The method of stochastic linearization provided a tool for reducing the nonlinear systems addressed in this volume to quasilinear ones.
2. The methods of linear control theory could be applied, with some modifications, to the quasilinear systems mentioned above.
3. The resulting relationships and techniques, as it turns out, are quite similar to those of linear control theory.

What is left for future developments in the QLC area? Plethora of problems! To mention a few:

A. As far as the method is concerned, stochastic linearization has not been justified analytically, and only numerical/experimental evidence of its accuracy is available. We believe that, in fact, it can be justified (as rigorously as the method of harmonic balance/describing functions).

B. As far as the control-theoretic issues are concerned, this volume addressed only odd nonlinearities. Non-odd ones are also encountered in practice. Therefore, one of the fruitful directions for future research is the development of a quasilinear control theory for feedback systems with non-odd instrumentation.

C. Today, practicing engineers apply, almost exclusively, the methods of linear control. It is an important task to make QLC methods available and conveniently applicable in practice. This could be enabled by developing a user friendly toolbox, which would place the QLC methods literally at the designer's fingertips.

We have no doubts that these open problems will be "closed" in our or other colleagues work. This, undoubtedly, would lead to other problems in the never ending and rejuvenating process of scientific and engineering discovery.

Abbreviations and Notations

Abbreviations

ILQR	instrumented LQR
ILQG	instrumented LQG
LC	linear control
LHS	left-hand side
LMI	linear matrix inequality
LPNI	linear plant/nonlinear instrumentation
LQG	linear quadratic Gaussian
LQR	linear quadratic regulator
NRRO	non-repeatable runout
QLC	quasilinear control
RHS	right-hand side
RL	root locus
RRO	repeatable runout
RS	random sensitivity
S	saturated
SLQG	saturated LQG
SLQR	saturated LQR
SRL	saturated RL
SRS	saturated random sensitivity
TD	trackable domain
wss	wide sense stationary

Notations

$1(t)$	unit step function
diag	diagonal matrix
$D(r)$	disk of radius r

dz_Δ	deadzone function
E	expectation
e_{ss}^{par}	steady state error with respect to parabolla
e_{ss}^{ramp}	steady state error with respect to ramp
e_{ss}^{step}	steady state error with respect to step
erf	error function
fri	friction function
I_0, I_1, I_2, I_3	quality indicators
k_s, k_s^+	system types
M_r	resonance peak
N	quasilinear gain
qz_Δ	quantization function
$\mathbf{R}$	Euclidean space
$RS(\Omega)$	random sensitivity function
R_{dc}	random d.c. gain
$R\Omega_{BW}$	random 3dB bandwidth
RM_r	random resonance peak
$R\Omega_r$	random resonance frequency
rel_α	relay function
sat_α	saturation function
sat	standard saturation function ($\alpha = 1$)
$SRS(\Omega)$	saturated random sensitivity function
SR_{dc}	saturated random d.c. gain
$SR\Omega_{BW}$	saturated random 3dB bandwidth
SRM_r	saturated random resonance peak
$SR\Omega_r$	saturated random resonance frequency
$S(s)$	sensitivity function
$sign$	sign function
tr	trace of a matrix
$u_\Delta(t)$	slanted step function
w	white Gaussian process
Σ_{zz}	covariance matrix
σ_x	standard deviation of x
ζ	damping ratio

Ω	3dB bandwidth
ω_{BW}	3dB bandwidth
ω_r	resonance peak frequency
ω_n	natural frequency
$\square$	SRL termination point
$\blacksquare$	SRL truncation point
$\|\cdot\|_2$	2-norm
$\|\cdot\|_\infty$	∞-norm

Index

Printed in the United States
By Bookmasters